PRODUCTION AND CHARACTERIZATION OF ALKALINE PECTINASE FROM *Bacillus tropicus* MCCC 1A01406

Dr. Pitambri Thakur

Associate Professor

University Institute of Biotechnology

Chandigarh University

TABLE OF CONTENT

Contents
  2-7

Abstract
  8-10

CHAPTER 1: INTRODUCTION
  11-17

CHAPTER 2: REVIEW OF LITERATURE
  18-49

2.1 Pectic substances

2.2 Pectin degrading enzymes

2.3 Sources of microbial pectinases

2.4 Optimization of pectinase production

    2.4.1 Submerged fermentation (SmF)

    2.4.2 Solid-state fermentation (SSF)

2.5 Purification of pectinases

2.6 Industrial applications of Pectinases

2.10 Biotechnological degumming process

2.11   Industrial production and commercial suppliers of Pectinases

CHAPTER 3: RESEARCH METHODOLOGY   50-81

3.1   Sample collection

3.2   Enrichment of alkaline pectinase producing microorganisms

3.3   Screening of alkaline pectinase producing bacteria

    3.3.1   Composition of Yeast Extract pectin (YEP) medium

    3.3.2   Sample processing

3.4   Screening to produce other hydrolytic enzymes by alkaline pectinase producing isolates

3.5   Pectinase assay

    3.5.1   Preparation of DNSA (Dinitrosalicylic acid) reagent

    3.5.2   Preparation of other reagents

    3.5.3   Assay procedure

3.6   Protein content estimation

    3.6.1   Reagents

    3.6.2   Estimation procedure

3.7   Selection of promising isolates for detailed study

    3.7.1   Localization of pectinase

    3.7.2   Optimum pH for activity of crude pectinase

3.7.3 Optimum temperature for activity of crude pectinase

3.8 Identification and taxonomical classification of isolate P3

    3.8.1 Growth characteristics

    3.8.2 Physiological and biochemical characteristics

    3.8.3 Molecular Characterization of alkaline pectinase producing bacterial strain

3.9 Effect of environmental parameters on pectinase production by *Bacillus tropicus* submerged and solid-state fermentation conditions

    3.9.1 Standard inoculum preparation

    3.9.2 Submerged fermentation (SmF)

    3.9.3 Solid state fermentation (SSF)

3.10 Purification of pectinase of *Bacillus* sp. P3

    3.10.1 Preparation of cell free supernatant

    3.10.2 Ammonium sulphate precipitation of the cell free supernatant

    3.10.3 Dialysis

    3.10.4 Concentration of dialyzed fraction

    3.10.5 Purification

        3.10.5.1 Ion exchange chromatography

        3.10.5.2 Gel filtration chromatography

    3.10.6 Sodium dodecyl sulphate-polyacrylamide gel electrophoresis (SDS-PAGE)

3.11 Characterization of purified pectinase produced by *Bacillus* sp. P3.

    3.11.1 Effect of pH on pectinase activity

    3.11.2 pH stability profile of pectinase

    3.11.3 Effect of temperature on pectinase activity

    3.11.4 Thermostability profile of pectinase

    3.11.5 Effect of chelating agents and surfactants

    3.11.6 Effect of various metal ions on pectinase activity

    3.11.7 Type of pectinolytic activity

    3.11.8 Determination of Michaelis-Menten constant (K_m) and V_{max} values

3.12 Application of pectinase produced by *Bacillus* sp. P3 in Degumming of buel (*Grewia optiva*) bast fibers

CHAPTER 4: RESULTS
63-129

4.1 Isolation and screening of alkaline pectinase producing bacterial strain

4.2 Gram character, shape, and localization of pectinase from different isolates

4.3 Effect of pH and temperature on the pectinase activity of bacterial isolates

4.4 Identification and taxonomical classification of isolate P3

    4.4.1 Morphological and Biochemical Characteristics of the isolate

    4.4.2 Molecular Characterization of the isolate P3

4.5 Effect of environmental parameters on pectinase production in submerged and solid-state fermentation systems

    4.5.1 Submerged fermentation

    4.5.2 Solid-state fermentation (SSF)

4.6 Purification of pectinase produced by *Bacillus* sp. P3

4.7 SDS-PAGE of purified pectinase

4.8 Characterization of purified pectinase produced by *Bacillus* sp. P3

    4.8.1 Effect of pH on activity and stability of the enzyme

    4.8.2 Temperature optima for enzyme action and thermal stability

    4.8.3 Effect of chemical agents and surfactants on purified pectinase activity

    4.8.4 Effect of metal ions on pectinase activity

    4.8.5 Type of pectinase activity

    4.8.6 Determination of K_m and V_{max}

4.9 Degumming of buel (*Grewia optiva*) bast fibers by RSM

    4.9.1 Bacterial treatment

    4.9.2 Optimization of enzyme dose and reaction time for degumming of fibers

4.9.2.1 Chemical Treatment

4.9.2.2 Experimental Design

4.9.2.3 Statistical analysis

4.9.2.4 Analytical assessment of degumming of fibers

4.9.2.5 Statistical analysis and model fitting

4.9.2.6 Validation of Regression model

CHAPTER 5: DISCUSSION
130-108

CHAPTER 6: SUMMARY AND CONCLUSIONS
109-138

REFERENCES
139-175

ABSTRACT

Alkaline pectinase is the utmost significant industrial enzyme of the bio scouring process. By considering bio scouring of cotton, 30 microbial isolates from fruit and vegetable waste-rich dump soil of Solang Valley and Vasishta (Manali, Himachal Pradesh, India) were isolated and screened for the alkaline pectinase production in the current research work. Only four isolates P3, P16, P21, and P27 were capable to produce extracellular alkaline pectinase at pH 9. Further by applying submerged fermentation, the alkaline pectinase production was quantitatively screened. The most efficient isolate was P3 identified as *Bacillus tropicus*, based on morphological, biochemical, and molecular characterization. Molecular characteristics confirmed by 16S rDNA sequence analysis. The nucleotide sequence of the isolate was novel with a 97% similarity index and submitted to the GenBank with accession number MK332379. This isolate was characterized as **Bacillus tropicus** MCCC 1A01406 based on morphological, biochemical, and molecular analysis. It was found to be Gram-positive, rod-shaped (2.0-2.5 pm), spore-forming, aerobic, non-motile, catalase, and oxidase-positive bacterium. **Bacillus tropicus** MCCC 1A01406 also produced other hydrolytic enzymes such as lipase, amylase, xylanase cellulase, and protease but lacked chitinase and tannase. Pectinase produced by **Bacillus tropicus** MCCC 1A01406 was extracellular. This strain grew well on most of the carbon sources used but produced acid only in glucose, fructose, sucrose, galactose, and maltose. The isolate grew well between pH and temperature range of 5.0-11.0 and 25-45°C respectively. This isolate was capable of hydrolyzing tributyine, xylan, starch, casein, and carboxymethyl cellulose, but did not hydrolyze tannic acid, chitin, and urea. Of the various complex media tested, yeast extract (1%, w/v) supplemented with pectin (0.25%, w/v), was found to support maximum pectinase production. Production of this enzyme was inducible. The optimum conditions for pectinase production by **Bacillus tropicus** MCCC 1A01406 were 37°C, pH 9.0, with shaking (250 rpm) for 16 h (with 2% v/v, inoculum size) in YEP medium. To enhance pectinase production in liquid YEP medium, various salts, carbon, and nitrogen sources were supplemented in the medium. The incorporation of $CaCl_2.2H_2O$ and $MgSO_4.7H_2O$ (1.0 mM) enhanced pectinase production almost three times. The addition of $CuCl_2.2H_2O$, $CoCl_2.2H_2O$, and $MnSO_4.4H_2O$ (final concentration, 1 mM) also enhanced pectinase production. Out of the various carbon, sources studied, mannitol (0.5%, w/v) enhanced pectinase production by almost 0.3 times whereas, most of the other carbon sources used except sodium acetate and polygalacturonic acid, inhibited the enzyme production. None of the different nitrogen sources (except yeast extract in the YEP) tested were found to enhance pectinase production.

Maximum pectinase production (55.7 U/ml) was obtained in yeast extract pectin medium supplemented with 1 mM $CaCl2.2H2O$ alone as none of the other optimized components alone or in combinations enhanced pectinase production as compared to $CaCl2.2H2O$. The production of pectinase was scaled up by solid-state fermentation. In SSF, using wheat bran as the prime solid substrate, 4600 U/g dry substrates of pectinase were obtained at 75% moisture content. Pectinase from **Bacillus tropicus** MCCC 1A01406 was purified to homogeneity by ammonium sulfate precipitation, ion-exchange chromatography, and gel filtration chromatography. Using DEAE Sephacel chromatography, 68.2 fold purification of the enzyme was achieved and its specific activity was found to be 738.6 U/mg protein. Further purification of this enzyme using Sephadex G-100, yielded 132.4-fold purification and the specific activity of the purified enzyme was 1466.2 U/mg protein. SDS-PAGE on 10% gel revealed a single band of the size 98 kDa. The purified pectinase was optimally active at 60°C and stable at 45°C for more than 4 h. The pH optima of the purified enzyme were 9.0 at 60°C and the enzyme was stable at pH 9.0, room temperature for more than 4 h. Mercaptoethanol (1 mM) stimulated enzyme activity by 45% whereas, urea, ascorbic acid, glycine, cysteine, and EDTA inhibited the pectinase activity. The surface-active agents such as tweens (80, 60, 40, and 20), triton-x-100, and SDS stimulated pectinase activity from 7 to 26% whereas, iodoacetic acid completely inhibited pectinase activity. The Ca ions stimulated enzyme activity by 44% however, Ag^{2+}, Cu^{2+}, Mn^{2+}, Zn^{2+}, Pb^{2+}, Ba^{2+} and Hg^{2+} inhibited pectinase activity. The pectinase produced by **Bacillus tropicus** MCCC 1A01406 seemed to belong to the pectic lyase family. The purified enzyme exhibited Km and Vmax values of 4.8 mg/ml and 390 U ml' min' respectively. **Bacillus tropicus** MCCC 1A01406, as well as crude pectinase from this bacterium, was used for the degumming of buel fiber crops *Bacillus tropicus* alkaline pectinase was used as a catalyst in the synthesis of galacturonic acid from beul (*Grewia Optiva*) bast fibers. RSM was used to optimize the reaction parameters for the synthesis of galacturonic acid and a second-order response model was calculated. The highest yield of galacturonic acid was 485mmole/g. The optimum reaction conditions for galacturonic acid synthesis were obtained as incubation time of 1.25 h, reaction pH of 8, alkaline pectinase concentration of 2g/l, and temperature of 42 °C.

CHAPTER 1: INTRODUCTION

All over the world environmental worry for waste disposal is same especially in the progressing countries [1-2]. Due to industrial and agricultural everyday actions, plenty of lignocellulosic waste like rice husk, sugarcane bagasse, etc. are released, causing a major disposal complication. This lignocellulosic waste has a massive volume of pectin, cellulose, and hemicellulose, later recycled to produce pectinase, cellulase, and hemicellulase respectively [3].

Pectinase is a group of one of the commercially important enzymes degrading pectin present in the plant cell wall. Pectinases degrade polysaccharide pectin to monogalacturonic acids and based on activity seven different classes of pectinases are pectinesterase (EC3.1.1.11), polygalcturonase (EC3.2.1.15), galacturan 1, 4-α-galactouronidase (EC3.2.1.67), exopoly-α-galactouronosidase (EC3.2.1.82), endopectate lyase (EC4.2.2.2), exopectate lyase (EC 4.2.2.9) and endopectin lyase (EC4.2.2.10) [4].

Pectic substances are polysaccharides with α (1-4) linkage in galacturonic acid backbone present as a cementing material in the middle lamella and the primary cell wall of all the plants [5,3]. The highest concentrations of pectin are found in the middle lamella of cell wall, with a gradual decrease as moving through the primary wall toward the plasma membrane (Figure 1.1) [6]. Pectin is a family of galacturonic acid-rich polysaccharides including homogalacturonan, rhamnogalacturonan I, and the substituted galacturonans rhamnogalacturonan II (RG-II) and xylogalacturonan (XGA). Although pectin occurs commonly in most of the plant tissues, the number of sources that may be used for the commercial manufacture of pectin is limited. This is because; the ability of pectin to form gel depends on the molecular size and degree of esterification (DE) (Figure 1.2).

At present, commercial pectins are almost exclusively derived from citrus peel or apple pomace, both of which are by-products from juice manufacturing units. Apple pomace contains 10-15% of pectin on a dry matter basis. Citrus peel contains relatively higher i.e. 20-30% of pectin as compared to that of apple. From an application point of view, citrus and apple pectins are largely equivalent [7]. Among the physical properties, citrus pectins are light cream or light tan in colour whereas apple pectins are often darker. Alternative sources for pectin extraction include sugarbeet waste obtained from sugar manufacturing, sunflower heads (seeds used for edible oil), and mango waste [8].

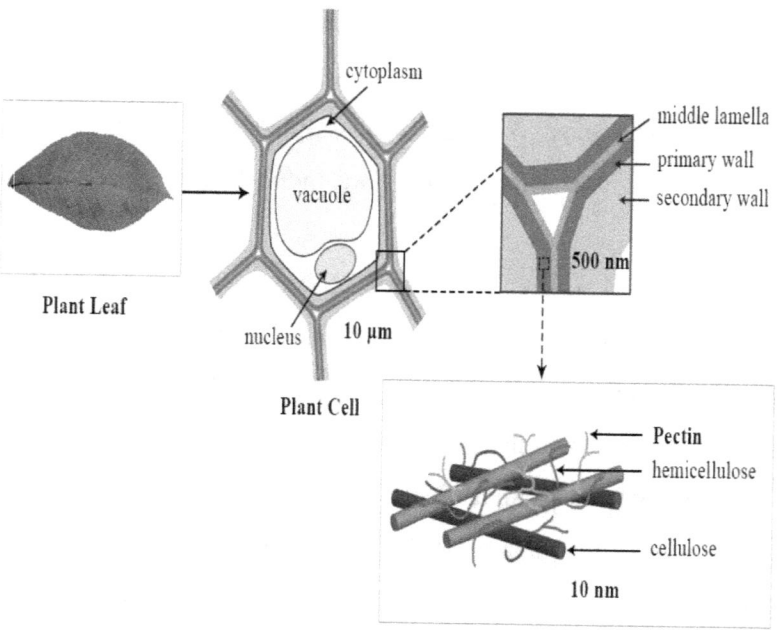

Figure 1.1 Schematic diagram showing location of pectin in the plant cell wall [6]

Apples, quince, plums, gooseberries, oranges and other citrus fruits contain much more pectin as compared to soft fruits like cherries, grapes and strawberries contain little pectin1 . Typical levels of pectin in plants are (fresh weight): Apples, 1-1.5; Apricot, 1; Cherries, 0.4; Oranges, 0.5-3.5; Carrots, approx. 1.4 and Citrus peels, 30%.

Industrial enzyme's global market is growing day by day and in the world market, the pectinases production report about 10% of the total enzyme production. By 2021 the market is estimated to reach around 35.5 million dollars [9]. In altering enzyme properties pH has a definite role. Acidic pectinases find a wide range of utilization in the fruit preparation industry for fruit juice clarification [10] and liquefaction of fruit juices [11] while alkaline pectinases are utilized in various industrial preparations such as fabric, pulp, and paper industry [12]. Alkaline pectinase is an emerging enzyme of commerce with primary employment in the paper and textile industries. Microorganisms are the main source of enzymes due to the usage of low-cost substrates [13].

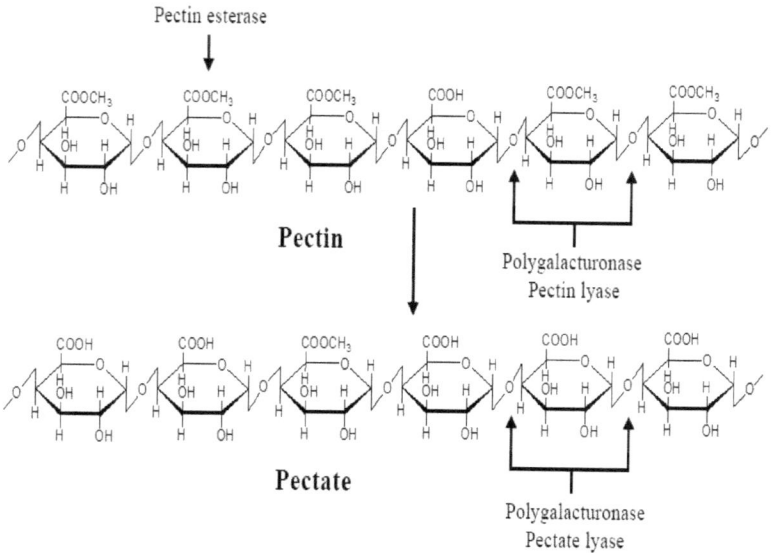

Figure 1.2 Structure of pectin and pectate as well as mode of action of pectine esterase, polygalacturonase and pectin lyase [7]

The textile industry is the utmost polluting industry due to maximum chemical usage. Traditional scouring involves alkaline chemicals to expel non-cellulosic material for soft and hydrophilic fiber, suitable for industrial applications. Due to the usage of alkaline chemicals textile industries releases wastewater with high values of total dissolved solids, chemical oxygen demand, and biological oxygen demand [14-15]. Thus, the need of the hour for sustainable development is the replacement of chemical processes with eco-friendly biochemical processes. Bio scouring employs enzymes to remove impurities without disturbing fiber's structure, strength, and environment [16].

Microbially derived pectinases are more in use over the plant and animal-derived pectinases due to low-cost production, faster product recovery, and easy gene manipulations. The main microbial sources of pectinolytic enzymes are yeast [17], bacteria [18], and a large diversity of fungi and particularly *Aspergillus* species [19]. Bacteria with well-characterized biology are well exploited as a source of industrially important enzymes. Bacterial pectinases are majorly extracellular enzymes with wide applications in bleaching of papers, wastewater treatment, and coffee fermentation [20-23]. One of the largest genera *Bacillus* from the *Firmicutes* family covers a great diversity of strains. *Bacillus* is endospore-forming, gram-positive, rod shape motile bacteria; some are aerobes or facultative anaerobes. Spores are formed under the adverse condition of nutrition or temperature or both. Low-cost production with new microbial isolates is fascinating the attention of many researchers nowadays [24]. Alkaline pectinase increases viscosity, breaking length, and porosity of the pulp [25]. Alkaline pectinases also lessen scouring chemicals in theplant fibers scouring.

Although acidic pectinases are one of the first enzymes utilized at homes and itseconomic utilization was first detected for wines and fruit juicesformulation in 1930. Since then, they have become an integral component, along with other hydrolytic enzymes, for the formation of good quality fruit juices at the profitable level. In the 1960s with an increased understanding of the molecular nature of the plant tissues, scientists useda wide range of enzymes more precisely [26-27]. Therefore, pectinases are one of the most utilized enzymes in the trading sector today.

Lately, alkaline pectinases have made exceedinglypromisinginvasion in the degumming as well as retting of fiber for thecreation of good quality textile material, and papermaking. Plant fibers are exceptional textile fibers due to the extensive strength and rigidity provided by thick-walled, lignified sclerenchyma cells of the plant [9].Conventionally plant fibers areretrieved either bythe mechanical processing of the bast from the plantor water retting of the stem. In both cases, bast fibers are liberatedfrom the surrounding tissues by maceration of the cortex, which is carried out by a diverseculture of pectinolytic organisms introduced as part of the normal natural flora in combinationwith the fibers.

Pawar et al., (2002) [28] suggested bio scouring is very easy to handle as compared to the alkaline traditional scouring process because of the presence of harmful chemicals. Bioscouring is a better process because scouring is done effectively without affecting fabric and the environment negatively. Bioscouring includesthe use of enzymes for the elimination of the non-cellulosic impurities and makes fibers more hydrophilic. Thus, as compared to alkaline traditional scouring, bioscouring is eco-friendlier and more cost-effective.

Thus, nowadays traditional pre-treatment of cotton processing can be easily replaced by an enzyme-catalyzed process. According to Sawada et al., (2003) [29],the backboneand sidechains of pectin are attached to the proteins and waxes. Pectin acts as cementing material that acts as a binder between the cell wall layers and present in abundance in the middle lamella. Experimental results for the comparison of chemical scouring with bioscouring show a clear-cut comparison between the two. For this comparison fibers were treated with chemical scouring and bioscouring separately then both were commonly bleached. Bioscouring presented better results with 55.65 whiteness index, 25.64 yellowness, and 27.74 brightness index, and 1.84g sugar/100g fiber [30]. Hence enzymatic scouring treatment proved to be eco-friendly and cost-effective for sustainable development.

The plant fibers are made up of long, narrow, thick-walled, and lignified sclerenchyma cells, contributingtoughness and rigidity to the plant, and are distinguished textile cloths. Formallyplant fibers areattained either by decorticationor water retting of the stem. In both cases, bast fibers are discharged from the surrounding tissues by maceration of the cortex, which is carried out by diverse populations of pectinolytic microorganismsimported as an element of the normal natural flora in combination with the fibers. This process takes a lot of time tocomplete and is weather-dependent. Moreover, the processed fibers still contain a large amount (20-40%) of gummy material consisting of pectin, cellulose, and hemicellulose. On an industrial scale, this gummy material is removed by treating the fibers with NaOH solution (12-20%) containing wetting and unwanted gummy material. The prolonged incubation of the fibers during water retting and the subsequent treatment with high concentrations of NaOH leads to consumption of time, damage to the fibers, and pollution problems. Therefore, to achieve 100% removal of gummy material, the existing technology must be modified, e.g., by use of monocultures instead of mixed consortia, the enzyme produced by these microorganisms, or asthe combination of the chemical and enzymes using lower concentrations of chemicals and to be more stable in alkaline conditions. Keeping in view the above, the following objectives were assessed for the present research work:

> ➤ Isolation of alkalophilic microorganisms from different environmental samples.
> ➤ Screening, selection, and characterization of alkaline pectinase production by the selected strain.

- Optimization of process parameters for the growth and pectinase production by the selected strain.
- Purification and characterization of the alkaline pectinase.
- Degumming of buel (*Grewia Optiva*) bast fibers using alkaline pectinase with the RSM

CHAPTER 2: REVIEW OF LITERATURE

All enzymatic mechanisms for instance the making of cheese, vinegar, and wine, the leavening of bread, and the brewing of beer have been the origin in prehistory. The enzymes have the advantage that the process becomes not only highly specific and quick but is also eco-friendly as it replaces chemical reactions, which invariably enhance environmental pollution. Due to the growth in the area of biotechnology, newer applications of industrial enzymes are emerging day by day and it is expected these will be used in every catalyzed factory process and in every home in the years to come. Enzymes are produced industrially because of their multifarious use in the dairy (coagulants) and detergents industries and collectively account for approximately 40% of all enzyme sales [31]. Carbohydrates, used in baking, brewing, distilling, starch, and textiles represent the second largest group. There are currently 12 main sectors under the "other" category, which include alcohol, animal feed, baking, chemical biotransformation, diagnostics, flavor, leather, textile, fruit and wine, paper and pulp, and waste (Figure 2.1).

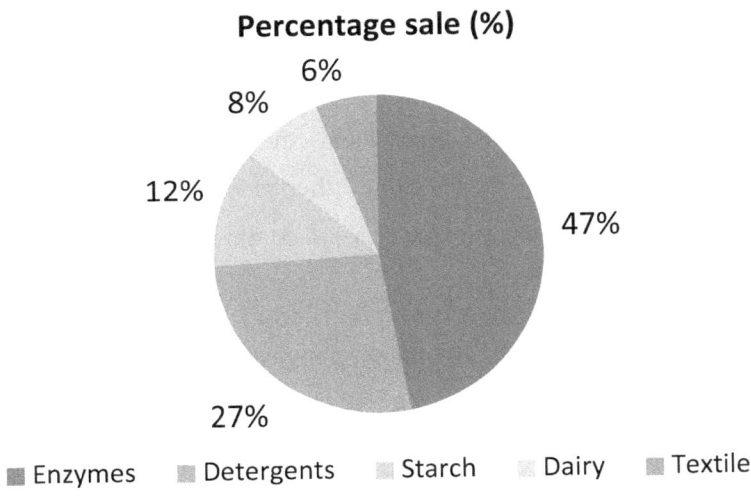

Figure 2.1: Outlook distribution of enzyme sales up to 2021

Pectinase is a group of the commercially important enzymes degrading pectin present in the plant cell wall. Pectinases degrade polysaccharide pectin to monogalacturonic acids and based on activity seven different classes of pectinases are pectinesterase (EC3.1.1.11), polygalcturonase (EC3.2.1.15), galacturan 1, 4-α-galactouronidase (EC3.2.1.67), exopoly-α-galactouronosidase(EC3.2.1.82), endopectate lyase (EC4.2.2.2), exopectate lyase (EC 4.2.2.9) and endopectin lyase (EC4.2.2.10) [32].

The enzymes have the advantage that the process becomes not only highly specific and quick but is also eco-friendly as it replaces chemical reactions, which invariably enhance environmental pollution. Due to biotechnology growth, newer applications of industrial enzymes are emerging day by day and it is expected these will be used in every catalyzed factory process and in every home in the years to come. Industrial enzymes have become an efficient means for processing agricultural products and bringing better value to food and beverage products. These are being used in a wide range of foods like pop drinks, wine, cheese,fruit juices, oils, beer, and meat to improve gains, processibility, mean life, taste,and other properties In 1983 there were some 30 different enzyme types in common commercial use.

Newer numbers are being added to this list every day. Proteinases prevailas the dominant enzymes produced industrially due to their multifarious usage in the dairy (coagulants) and detergents industries and collectively account for approximately 40% of all enzyme sales [33]. Carbohydratesutilizedinbaking, brewing, distilling, starch, and textiles represent the second largest group. There are currently 12 main sectors under the "other" category, which include alcohol, animal feed, baking, chemical biotransformations, diagnostics, fats and oils, flavor, fruit and wine, leather, textile, paper, and pulp and waste. The growth of these sectors in the coming years is expected to be such that the distribution of sales for 2021 will be around 50% of total enzyme sales. It is expected that the "other" sector will collectively be the single largest part of enzyme demand [34].

Pectinases represent a class of hydrolytic enzymes that are subjected to the degradation of long and complex polysaccharides, termed pectic substances, which are present as structural compounds in the cell walls of young plant cells.

2.1 Pectic Substances

In the past, several terms have been suggested for distinct pectic substances. However, in 1944, the Committee for the Revision of Nomenclature of Pectic molecules defined "Pectic Substances" as a class of plant derivatives containing a huge fraction of anhydrogalacturonic acid units, which occur as chain-like aggregates.

2.1.1 Occurrence and significance

Pectic polysaccharides occur in the middle lamella and are found as a gummy "cementing" substance in the young plant's primary cell walls [35]. Plant tissues are composed mostly of a relatively unspecialized tissue called parenchyma. The parenchyma cells have thin walls composed of two layers. Apart from functioning as a lubricant agent in the cell walls, pectic materials also create synergy between plant roots and growth, ripening, and storage [36]. The quality of fruits and vegetables is greatly determined by the presence of pectin type. The most indicative change during the ripening of fleshy fruits is softening. In the green, immature fruit, the pectin is attached to the cellulose microfibrils and is insoluble, therefore provide rigidity on the cells.In the course of ripening, these pectic substances are degraded and solubilized [37-38] leading to softening of the fruit.

2.1.2 Nomenclature

Depending upon the type of alteration of the backbone chain, Committee for the Revision of the Nomenclature of Pectic Substances, in 1944 pectic substances were classified into four main categories [39].

2.1.2.1 Pectins

The general term "pectin" (or pectins) refers to that water-soluble polymeric molecule containing most 75% of the carboxyl molecules of the galacturonic acid are esterified with methanol and under suitable conditions is proficient in forming gels with sugar and acid [40]. For this reason, two groups of jellying pectic substances are distinguished: the high-methoxylpectins (>50%) and the low methoxylpectins. The term "high-methoxylpectins" is often abridged to "pectins".

2.1.2.2 Pectinic acids

The colloidal polygalacturonic acids with methyl ester groups are called pectinic acids. Pectinic acid displays different properties of gel formation with sugar if appropriately low in methyl composition.

2.1.2.3 Pectic acids

Pectic acids are denominated in pectic materials usually made up of colloidal polygalacturonicacids, and are typically free from methyl ester groups [41]. Most of the pectic acids are formed after the dissolution of the tissue, via the action of pectin methylesterase.

2.1.2.4 Protopectin

Protopectin is an original pectic molecule and produces pectin on restricted breakdown. Protopectin is periodically called a water-insoluble pectic substance constituted and produces water-soluble pectic molecules [42]. The insolubility of the protopectin can be a function of the polymer size as 9+ well as the presence of divalent cations such as Ca.

2.1.3 Chemical constitution and structure

Pectic molecules contain a rhamnogalacturonan backbone with a "smooth" a-D-1, 4- galacturonan domain which is disrupted to a limited expansion by the introduction of 1,2-linked a-L-rhamnosyl molecules and exceedingly divergent blocks with an approximately alternative rhamnogalacturonan chain [43-44]. Sidechains composed of neutral sugars are attached by glycosidic linkages to carbon atoms 3 and 4 of the rhamnose units and 2 and 3 of the galacturonic acid units, producing the rhamnogalacturonan portion of the pectin a "hairy" character [45]. The prevalent sugars D-galactose and L-rhamnose are present in complex chains of appropriate length [46], whereas some xylose occurs in monomeric or oligomeric side chains. Acetic acid may occur as a substitute on the hydroxyl groups of galacturonic acid at carbon atom 2 or 3[47]. The side chains link the pectin molecules to proteins, hemicellulose, and cellulose to form the insoluble protopectin [48].

2.1.4 Properties

A pectic substance's solubility in water depends on the length of the pectin molecule. It increments with the lesser length of the pectin molecule and vice versa. Final esterification of the carboxyl groups (-COOCH3, -COOCH2CH2OH) or the partial substitution at the secondary hydroxyl groups (-OCOCH3, -OCH2OH) induces water solubility. The aqueous solution (1-2%, w/v) of pectic substances having high molecular weight is viscous. The viscosity of these substances increases with higher molecular weight and a higher degree of esterification and is also influenced by ionic strength, pH, and temperature [49]. In acid solutions, the degree of esterification and polymerization decreases. In alkaline solutions and at low temperatures, saponification of the methyl ester groups occurs readily.

However, depolymerization is strongly enhanced by a rise in temperature. Pectin powder absorbs water in a humid atmosphere but dissolves exclusively in a surplus of water. Insoluble pectinates, cross-linked pectins, and protopectins displaydelimited swelling in water. Chelating agents, such as 2+ ethylenediaminetetraacetate increases the swelling of protopectin, which removes Ca and other polyvalent cations. The most uncommon and exclusive physical properties of pectins are their ability of gel formation with sugar and acids which are thermo-reversible [50].

2.1.5 Biosynthesis of pectic substances

The synthesis of polysaccharides has been studied by easy incorporation of radioactive carbon into the leaf and fruit pectin of plants grown in anatmosphere containing C1402. Based on these studies, glucose, and galactose has been considered as pectin precursors. Uridine phosphate (UDP) is possibly part of these reactions. Methionine is bound for the transfer of this methyl group to the pectin molecule [51].

2.1.6 Pectin manufacturing: Chemical extraction and purification

The raw material for pectin production is by-products during the manufacturing of fruit juices: apple pomace (dried) and citrus residues [52]. Until, recently, chemical extraction has been the only way to produce pectin. This extraction is performed by an acid breakdown in a pH range of 2.0-3.0 for 0.5-5 h at a temperature range of 70-100°C. And the solid to liquid ratio is normally about 1: 18. By using centrifugation or hydraulic press pectin is extract is removed from the pomace. Subsequently filtered, concentrated then employed organic solvents to precipitate, finally, precipitates are collected and dried [53].

2.1.7 Applications of pectin

2.1.7.1 In the food sector

Pectin is utilized as a jellifying functionary to give a gelled texture to foods, mainly fruit-based foods, especially jams and jellies [54]. Pectin is considered to be the best jellifying agent under a low pH. Pectin establishes a texture that retains a uniform distribution of fruit particles during transportation, gives a good flavor release, and minimizes syneresis [55]. The pectin concentration used varies from 0.1-0.4% in jams and jellies. In dairy products, the pectin reacts with casein, preventing the coagulation of the casein' at a pH below the isoelectric pH (4.6) and allowing pasteurization of the sour milk products to extend their shelf life. Also utilized for the single-cell protein production in a modified "Symba process" [45].

2.1.7.2 In the pharmaceutical sector

Pectin belongs to a divergent class of molecules referred to as "dietary fiber", possesses several valuable biological effects. It is known asan "intestinal regulator" and acts as a detoxifying agent [56]. Pectin reduces pharmaceuticals toxicity without lessening its therapeutic effect [57]. More recently, pectin gelation micro globules were developed extensively for intravascular biodegradable drug delivery systems [58].

2.1.7.3 In the cosmetic sector

The applications in the cosmetical sector utilize the "ordinary" properties of pectin. Examples are the numerous gels (hair) and pastes [58].

2.2 Pectin degrading enzymes

Pectin degrading enzymes are divided into two groups, depolymerizing enzymes and saponifying enzymes orpectinesterases (Figure 2.2). Depolymerizing enzymes are again classified into the three major types of pectinases discussed below:

2.2.1 Esterases

Pectinesterases (PE, EC 3.2.2.11) saponify the methyl ester groups of pectic substances by a nucleophilic attack. The enzyme works in preferencetothe methyl ester group of a galacturonate unit adjacent to the free carboxyl groups and then proceeds along the substrate in a block-wise distribution of carboxyl and esterified carboxyl groups [59]. However, some pectin esterases attack pectin at the reducing chain end, at the same time others attack the non-reducing end [60].

Esterases are active at a pH range between 4.0-8.0. Fungal origin pectinesterase has lower optimum pH as compared to bacterial origin [61]. Alkaline and acidic PEs demethylates pectic molecules in different patterns [62]. These enzymes are formed by fungi [63], bacteria [64], yeasts [65], actinomycetes and higher plants [40].

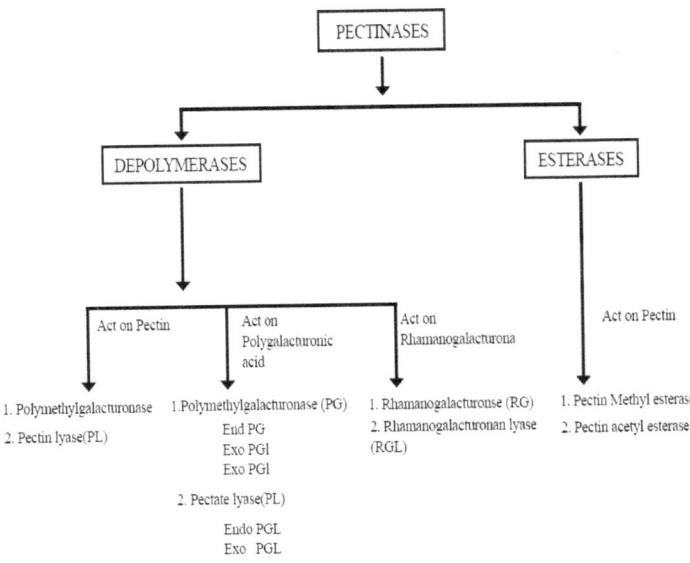

Figure 2.2 Pectinases classification based on mechanism of action on various pectin substrates [59]

2.2.2 Hydrolases

The enzymes belonging to this group act either on pectic acid (polygalacturonases) or pectin (polymethygalacturonases) by cleavage of glycosidic bonds with the aid of water. These two kinds of hydrolases are discussed below:

2.2.2.1 Polygalacturonases (PG)

Polygalacturonase (synonyms: pectinases, pectolases, pectin glycosidases, pectin depolymerases, etc. EC 3.2.1) are the hydrolytic enzymes. In general, by the action of PGs pectic acid is broken down into mono-, di-, and trigalacturonic acid. These end products may be produced by a "single-chain multiple attack" mechanism, they can be detected easily, or by a "multi-chain attack" mechanism, when mono-, di-, and trimers assemble only after further hydrolysis of the initial depolymerization products [64]. Endo-PG catalyzes random hydrolysis of a-1, 4 - glycosidic linkages in pectic acid, whereas Exo-PG (EC 3.2.1.67), activates the glycolytic splitting of digalactosiduronatesequentially [65, 66] giving galacturonic acid as the major reaction product.

Endo-PGs are optimally active at a rather low pH (4.0 to 6.0) and a temperature of 30-40°C [66]. Exo-PGs have pH optima from 4.0 to 6.0. Endo-PGs are produced by a variety of microorganisms, numerous fungi [67] yeasts [68], higher plants, and some plant parasites [69]. Exo-PG, however, has been shown to occur in different fruits and vegetables as well as in fungi and bacteria. Methods of the first group are based on the determination of reducing groups released because of substrate hydrolysis by PG, whereas the second group involves measurement of drop-in viscosity [70].

2.2.2.2 Polymethylgalacturonases (PMG)

Pectic enzymes preferentially degrading pectin are called polymethygalacturonases. These enzymes cause either random (endo-PMG) hydrolytic division of a-1,4- glycosidic bonds of pectinor a sequential (exo-PMG) hydrolysis ofα-1,4 glycosidic linkages [71].

Although some reports are describing the catalytic activities of PMGs [72], the appearance of these enzymes has been mistaken for polymethylgalacturonase-containing preparations. Assay methods like those applied for determining PG activity can also be applied here, with exception of the course of the substrate to be used.

2.2.3 Lyases

The enzymes belonging to this group of pectinolytic enzymes catalyze the breakdown of either pectic acid (polygalacturonate lyases) or pectin (polymethyl galacturonate lyases) by P-elimination reaction. The two kinds of lyases are discussed below:

2.2.3.1 Polygalacturonate lyases (PGL)

Lyases (or trans-eliminases) catalyze a non-hydrolytic p-elimination between pectates (endo-, EC 4.2.2.2 and exo-polygalacturonate lyase, EC 4.2.2.9) [73]. In some aspects, PGL can be distinguished from PG. Their pH optima are significantly higher, ranging from 8.0 to 10.0. Generally, pectates are good substrates for endo-PGL, but some enzymes exert optimal activity on pectin with a specific degree of polymerization [74]. The product from PGL is unsaturated galacturonic acid. Exo-PGLs release unsaturated oligogalcturonates from the reducing end of the polymer.

PGLs are predominantly of microbial origin. Their best-known producers are *Erwinia* and *Bacillus* [73], along with *Aeromonas* [74], *Pseudomonas* [75], *Xanthomonas* [76], *Aspergillus* [77], and *Fusarium* [78]. To detect lyase activity especially, one assay has been widely and frequently used: measuring the increase in light absorption by reaction mixtures at 230 or 235 nm. At this wavelength, the double bond produced on the tans-eliminative cleft of the substrate absorbs the most. The unsaturated di- and oligouronides also react with thiobarbituric acid to form red chromogen with maximum absorption at 545-550 nm [79].

2.2.3.2 Polymethylgalacturonate lyases (PMGL)

PMGLs catalyzeelimination among the non-reducing end of the pectinates (endo-polymethylgalacturonate lyase). These enzymes are found in some fungi, but rarely in bacteria [80]. All

PMGLs are endo-acting enzymes that cause a rapid drop in viscosity and have a pH range between 5.0-9.0.

2.2.4 Protopectinases

Depending on their active site, protopectinases have been classified into two types. Protopectinases react with the protopectin composed of methoxylated galacturonic acid molecules. These protopectinases are A-type protopectinases. This action of protopectin is brought about either by simple hydrolysis or by a trans-eliminative mechanism. And B-type reacts with rhamnogalacturonans and neutral sugar side chain [81]. Protopectinases are produced by several yeasts [82], bacteria [83], and fungi [84].

2.3 Sources of Microbial pectinases

Pectinases can be produced by diverse microorganisms like bacteria [85-88], yeasts [89], fungi [90-92], and actinomycetes [93]. Based on pH optima at which these enzymes act, pectinases are categorized into broad classes: acidic pectinases andalkalinepectinases. Table 2.1 contains the list of pectinases producers recorded in the literature by various researchers.

Table 2.1: Bacterial and fungal sources of pectinases.

Microorganisms	Pectinase Type	Activity opti. pH	Activity opti. temperature (°C)	Reference
Bacillus Stearothermophi	PGL	9.0	70	[85]

llus

Organism	Enzyme	pH	Temp	Ref
Bacillus GK-8	Exopectinase	8.5	-	[88]
Bacillus sp. NT-33	PG	10.5	75	[93]
Amicola sp.	PGL	10.25	70	[94]
Erwinia carotovora	PL	9.0	50-60	[95]
Bacillus subtilis Strain SOI 13	PL	8.4	42	[96]
Bacillus sp. MG-cp-2	PG	10.0	60	[97]
Lactobacillus plantarum Strain 11	PG	4.5	35	[98]
Geotrichum lactis	Exo-PG	5.0	40	[99]
Bacillus sp. RK9	PGL	10.0	-	[100]
Bacillus No. P-4N	PGL	10.0	60	[101]
Bacillus polymyxa	PG	8.4-9.4	45	[102]
Bacillus pumilis	PGL	8.0-8.5	60	[103]
Bacillus subtilis	PL	8.5	60-65	[104]
Trichoderma resii	PE	7.6	-	[105]
Bacillus sp. RK9	PGL	10.0	60	[106]
Bacillus subtilis	PGL	7.5-8.3	-	[107]
Bacillus polymyxa	PGL	8.3-9.6	-	[108]
Bacillus sp. NT-33	PG	10.5	75	[109]
Pseudomonas syringae pv. glycinea	PL	8.0	30-40	[110]
Aspergillus niger CH4	Endo-pectinase, Exo-pectinase	4.5-6.0, 3.5-5.0	Below 50	[111]
Penicillium frequentans	Endo-PG	4.5-4.7	50	[112]

Sclerotium rolfsii	Endo-PG	3.5	55	[113]
Rhizoctonia solani	Endo-PG	4.8	50	[114]
Saccharomyces Sereviceae	PMGL	8.5	30	[115]
Penicillium frequentans	Endo-PG	4.8	50	[116]
Mucor pusilus	PG	5.0	40	[117]
Saccharomyces Cerevisiae	PG PE	5.5	30	[118]
Penicillium italicum CECT 22941	PL	8.0	50	[119]
Aspergillus niger	Endo-PG Exo-PG	4.0-4.9	-	[120]
Rhizopus sp. LKN	Endo-PG	4.5-4.7	55-60	[121]
Pseudomonas syringae pv. *Glycinea*	PL	8.0	30-40	[122]
Aspergillus niger	Exo-PG Endo-PG	4.5	45	[123]
Sclerotinia sclerotiorum	Exo-PMG	5.0	45	[124]
Aspergillus niger VTTD-77050	PG	4.5-5.0	50	[125]
Verticillium albo-atrum	End-PG	4.6-5.0	46	[126]
Fusarium oxysporium f. sp. *Melonis*	Exo-PG Endo-PG PGL	5.0	40	[127]
Cryptococcus albidus	Endo-PG	3.75	37	[128]
Penicillium italicum CECT 22941	PL	8.0	50	[95]
Pseudomonas Marginalis	PGL	8.5	-	[129]
Xanthomonas campestris	PGL	9.5	25-30	[97]

Neurospora crassa	PG	6.0	45	[129]
Erwinia carotovora	PL	9.0	50-60	[97]
Saccharomyces sereviceae	PMGL	8.5	30	[99]

2.4 Optimization of pectinase production

Many researchers have investigated various conditions of pectinase production by submerged fermentation (SmF) [130-131], and Pedrolliand Carmona (2010) [132] have correlated the pectinase yields and productivity, and Acuna-Arguelles et al (1995) [101] has correlatedthe enzyme profiles and kinetic features of different activities. Some of the culture condition reported in the literature for pectinase production by SmF and SSF include:

2.4.1 Submerged fermentation

2.4.1.1 Effect of medium pH

While studying the pectinase synthesis from different bacterial species the optimum pH has been recorded to vary from 7.0-10.0 [133-137]. However, in a few cases like *Xanthomonas campestris* slightly acidic pH (6.8) has been preferred [138]. Fungi and yeasts grow optimally at acidic pH [139-142]. The optimal pH for the production of pectinase by fungi and yeasts has beenrecorded in the acidic rangee.g. 3.7 for *Aspergillus* sp. [143], 3.0 for *Rhizopus nigricans* [144]; and 2.5 for *Penicillium frequentans* [145].

2.4.1.2 Effect of incubation temperature

Different literature records support the synthesis of pectinases from many bacteria and fungi at 30°C or less, but there are also reports on pectinase production at 37°C or even at higher temperatures. Different temperature optima reported in the literature for bacterial cultures include 27°C, for *Pseudomonas suringae. Glycinea* [146]; 25°C for *Aeromonas liquefacience* [147]; 30 °C for *Pseudomonas marginalis*[148]; *Bacillus* sp. KSM-P15 [149]; 37 °C for *Bacillus* No. P-4-N [150] and 65° C for *Bacillus stearothermophillus*[151]. Most fungi grow optimally at 30°C, but their growth and pectinase production at lower, as well as higher temperatures than this, has also been reported. *Fusarium oxysporium* f. sp. melonis has been reported to have optimal pectinase production at 26°C whereas *Penicillium italicum* has temperature optima of 28°C. Paradoxically, *Aspergillus* sp. CH-Y-1043, which has been intensively studied for commercial production of pectinases, grows and produces pectinase optimally at 37° C [152].

2.4.1.3 Effect of carbon supply

The supply of carbon in the growth medium is important for both cell growth as well as for enzyme production. Many researchers have reported the use of pectin as the important source of carbon for pectinase generation. PGA or pectin which initially acts as inducers, once acted upon by pectinases serves as a carbon source. Hsu and Vaughn (1969) [153] has reported hyperproduction of pectinase from *Aeromonas liquefaciens* by feeding glucose, glycerol, or PGA (2%), to carbon limited culture. PGA is the most effective carbon source for *Erwinia* sp. whereas *Yersinia enterocolitica* utilizes glycerol for the highest PATE production [154]. *Pseudomonas marginalis* has been reported to grow on 0.5% glucose, PGA, or pectin [155]. Alkalophilic *Bacillus* sp. RK9 utilizes 1% of sucrose or sodium pectate as a carbon source [156]. Another strain of *Bacillus* sp. KSM-P15 utilized pectin for pectate lyase production [157]. Alkalophilic *Bacillus* GK-8 has been reported to utilize 1% glucose to produce constitutive alkaline exopectinase [158]. In addition, several pectin-based waste products have also been reported to be used as carbon sources. Berbegal *et al*, (2017) [159] have aimed their work utilizing apple pomace for pectinase production. Aguilar and Huitron (1990) [90] studied the constitutive formulation of exopectinase by *Aspergillus* sp. CH-Y-1043 utilizing 0.5% of different carbon sources: citrus pectin, PGA, galacturonic acid, glucose, fructose, and glycerol. Pericin *et al* (1992) [160] have reported the use of 2% apple pomace, 0.5% sucrose, and 1% dry whey as the carbon source in submerged culture for pectinase production by *Polyporous squamous*. 5% glucose was used for endo-PG production by yeast *Kluyveromycesmarxianus*[161]. *Saccharomyces cerevisiae* has been reported to utilize 1% glucose, glycerol, ethanol, apple pectin, or galacturonic acid for PG production [162].

2.4.1.4 Effect of nitrogen supply

Nitrogen supply can broadly be provided as organic or inorganic supply. Yeast extract has been most widely utilized as an organic nitrogen source. It has also been reported to serve as an inducer [163]. Many scientists have used yeast extract in combination with an organic and inorganic source to produce pectinolytic enzymes. A combination of 0.2% yeast extract and 0.1% NH_4Cl has been reported by Member and Burlot (1994) [164] for *Pseudomonas marginallis* pectinase. Horikoshi (1990) [66] used 0.5% yeast extract and 0.5% polypeptide for alkaline pectinase generation by *Bacillus* sp. P-4-N. An alkaline *Bacillus* sp. KSM-P15 utilized 3% polypepetone and 0.5% yeast extract for maximal pectinase production [165]. Another alkalophilic *Bacillus* has been cultured in a medium consisting of 0.5% pectin and 0.5% peptone to produce alkaline exopectinase[166].

2.4.1.5 Effect of inducers

Pectinolytic enzyme production in some organisms is constitutive whereas in others it is inducible. Constitutive PATE (polygalacturonic acid transeliminase) production has been reported in the case of *Aeromonas liquefaciens*, but catabolic repression has been observed in the case of excess substrate [153]. Aguillar and Huitron (1990) [90] reported constitutive production of exo-pectinase. However, some organisms have an inducible enzyme system for pectinase production, pectinase being the best inducer followed by PGA [167].

2.4.1.6 Effect of salts

Calcium is outstandingly essential to produce many extracellular enzymes. Although PGL production in the absence of calcium, has been reported [168], but the level of enzyme is reported to increase considerably with increased levels of calcium. Horikoshi (1990) [66] has illustrated the enzyme of *Bacillus* No. P-4-N being stimulated by calcium. Other monovalent, as well as divalent cations, are reported to inhibit pectinase production alone or in the presence of calcium [169]. EDTA is acting as the pectinase production inhibitor [170-173].

2.4.2 Solid state fermentation(SSF)

The solid-state fermentation exhibits considerable advantages as compared to the submerged fermentation (SmF) process. These are (a) lower water requirement; (b) cheap media for the fermentation; (c) less stringent aseptic conditions; (d) utilization of huge concentration of substrate and (e) generation of huge product concentration [174-178]. Agro-industrial waste for example wheat bran, rice bran, sugarcane bagasse, corncobs, and apple pomace is mostly investigated as the best substrates for the

SSF process [179]. Pectinase production using wheat bran [180]; citrus pulp-pellets [181]; coffee pulp [182] and sugarcane bagasse [183-184] recorded in the literature. The most important factors for enzyme production are the substrate's molecular size and moisture level [185]. In SSF processes the galacturonase production decreases at low water activity values [186]. A comparative study has been conducted for the pectinase generation in SmF and SSF from *Aspergillus* sp., in which increased pectinase generation has been reported [187-188]. Pectinase produced by *A. niger* in SSF in comparison to those produced by SmF is more thermotolerant also has a broader range of pH values [189].

2.5 Purification of pectinases

Many mechanisms have been characterized in the previous findings for the purification of pectinases. These include (i) Non-specific procedures as precipitation techniques (salts, e.g., ammonium sulphate; organic solvents e.g., acetone and ethanol; organic polymers e.g., polyethylene glycol), gel filtration, ion-exchange chromatography, and (ii) Specific procedures, such as affinity chromatography [190]. Some of the bacterial pectinases reported in the literature, which have been purified, are indexed in table 2.2.

Table 2.2: Purification of the bacterial pectinases reported in the literature.

Producer	Type of enzyme	The method used for purification	Folds purification	Size (KDa)	Reference
Bacillus No.	PG	Sephadex G-160, DEAEcellulos	300	-	[120]

P-4-N		e followed by Sephadex G-200			
Bacillus pumilus	PATE	Ammonium sulphate precipitation, calcium phosphate gel adsorption and Sephadex G-100 column chromatography	16	20	[122]
Bacillus Polymyxa	PG	DEAE-cellulose	-	-	[121]
Bacillus stearothermophilus	PATE	Sepharose, polygalacturonamide linked matrix	62	24	[85]
Bacillus sp. KSM P-15	PL	Ammonium sulphate Precipitation	-	20.3	[191]
Xanthomonas Campestris	PATE	Acetone precipitation, citrate extraction, DEAEcellulose column chromatography	66	-	[123]
Closridium Multifermentans	Exo-PG PE	Sephadex G-120	156 178	-	[192]

2.6 Industrial applications of pectinases

With respect to industry acidic pectinases are utilized specifically to separate and clarify fruit juices [193-195], extraction of oils such as lemon oil [196], treatment of crushed grape seeds and juices for winemaking [197-198], tea and coffee fermentation [199] production of Sauerkrant juice by liquefaction of white cabbage and Sauerkrant[200] while alkaline pectinases are found extensively to be utilized for the retting and degumming of the fiber [201-205] and advance the treatment of fruit processing discharge [206]. Some important application of the pectinases is discussed below (Figure 2.3):

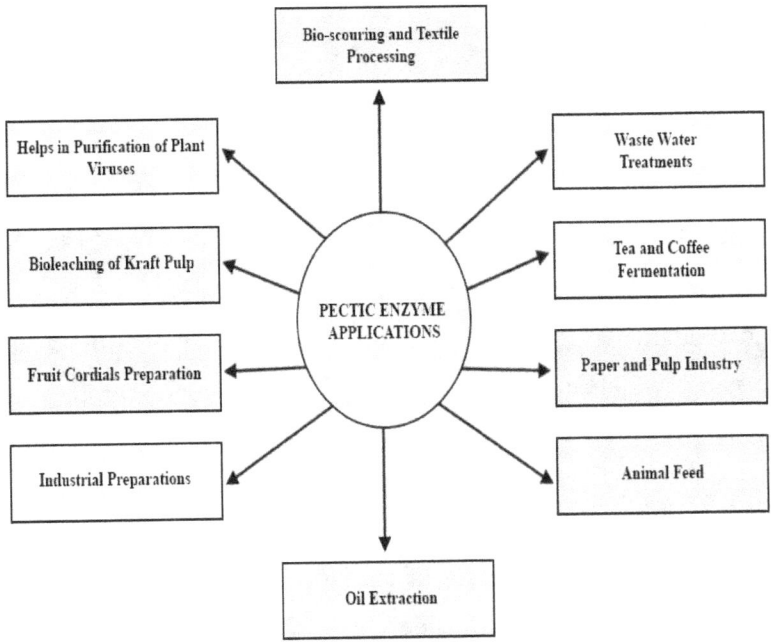

Figure 2.3: Various industrial and biotechnological applications of the pectic enzyme [207]

2.6.1 Acidic pectinases

Acidic pectinases are utilized in the fruit juice industries to produce: (A) Gleaming clear juices, and (B) Juices with clouds. Other applications of the acidic pectinases are in the preparation of ciders, wine, recovery of peel oils, preparation of dried animal feed, and coffee and tea fermentation.

2.6.1.1 Gleaming clear juices

To enhance the juice yield in the process of pressing and straining and to expel the haze enzymes are added to produce gleaming clear juices. After that juice is pasteurized to denature the leftover enzymes if a cloudy product is needed [208]. After centrifugation, all the debris is eliminated except some suspended molecules which are removed by utilization of the commercial dose of enzymes containing pectinases, cellulases, and hemicellulases [209-211].

Apple juice clarification could be achieved by mixing PG and PME without any infecting enzymes from the apples [212]. Research has also reported that pure pectin lyase can be utilized to clarify 91% -92% esterified apple juice at pH 3.0-4.0 [213-215]. The haze may be generated by the polymerization of polyphenols and oxidation of proanthocyanidinsin the course of milling and pressing, which can be removed by using pectinolytic enzymes [216-218].

2.6.1.2 Cloudy juices

For the sustainability of the cloud of purees, citrus juices, and nectars, pectinases are added [219-220]. Enzymatic treatment of orange juices has been reported to increase cloud stability. Pectic lyase with PME activity has been used successfully to produce these juices, as this enzyme does not attack the insoluble pectin which controls the sustainability of the cloud [221-222]. In the production of lemon juice, peels and pulp are ground, blended with water (1:1), heated at 95°C to eradicate the citrus PME, and cooled down at 50° C. To obtain a stable juice it is treated with pectinases [223-224].

2.6.1.3 Other applications

Other applications of acidic pectinases are described below:

2.6.1.3.1 French ciders production

For the production of the high-quality French ciders which are bittersweet in taste from the cider apples, the utilization of pectinase is required [225]. In this regard, apple juice production is the first step followed by the utilization of the trading fungal pectinases for clarification step [226-227]. Pectinase enzymes facilitate the production of high-quality cider by controlling and accelerating the mechanism of coagulum formation [228-230].

2.6.1.3.2 Wine making

In the winemaking pectinases from fungi, mostly *Aspergillus niger*, *Penicillium notatum*, or *Botrytis cinerea* are utilized. Grape wines are formulated in greater volume, despite this, juices of berries, peaches, apples, pears, and other fruits are also selected [231-234]. At first, to reduce pressing time pectinases are added to crushed grapes to increase the volume of free-run juice [235-237]. During the second stage at the time of fermentation enzyme can be added to settle out suspended particles, to increase filtration rate and wine clarity.

2.6.1.3.3 Oil extraction

Traditionally oils from different seeds such as coconut, sunflower seed, olives, and rapeseed are extracted with organic solvents [238-239]. These solvents are carcinogens. So, pectinases are utilized for the extraction of vegetable oil in an aqueous process.

2.6.1.3.4 Tea and coffee fermentation

For the fermentation of coffee and tea pectinases are the most important enzymes. Pectinolytic microorganisms are utilized to eradicate the mucilage coat of the coffee beans [240-241]. Sometimes pectinases are utilized for the separation of the bean pulpy layer and cellulases and hemicellulases provide additional aid for the digestion of the pulpy layer [242].

2.6.1.3.5 Preparation of the animal feed

Animal feed can be formulated by the utilization of peels, seed waste, and pulp. The dried pomace after enzyme treatment is utilized as a source of concentrated feed nutrient for sheep and dairy cattle. After juice extraction from the fruits, the oil-water emulsion remains entrapped in the peel particles [243].

2.6.2 Applications of the alkaline pectinases

Alkaline pectinases are mostly utilized for the degumming and retting of fiber, pectic wastewater pretreatment, and paper-making [244].

2.6.2.1 Retting and degumming of fiber crops

The plant fibers are classified into the following three main types

(i) Bast fibers
They are formed in groups outside the xylem in the cortex, phloem, or pericycle e.g., Ramie (*Boehmeria nivea*), sun hemp (*Crotolariajuncea*).

(ii) Structural fibers
These fibers are supportive and conducting fibrovascular bundles, chiefly present in monocots e.g., Manilla hemp (*Musa textilis*).

(iii) Surface fibers

These types of fibers are present on the surface of stems, leaves, etc. e.g., cotton (*Gossypium* spp).

Pectinases are utilized for the retting and degumming of flax, hemp, ramie, and coir from the coconut [245]. In the process of retting, many bacteria (e.g., *Clostridium, Bacillus*) and fungi (e.g., *Aspergillus, Penicillium*) are utilized to breakdown pectin of the stem to fiber [246].

2.6.2.1.1 Buel (*Grewia optiva*) bast fibers

Grewia optiva is a useful, medium-sized tree (7-10 m) with a spreading crown. In India, the tree is found from Kashmir to Sikkim, generally on the boundaries of cultivated fields. In Nepal, it is found in all hilly regions. In India, common local names for *Grewia optiva* are: "behul", "buel", "bhekua", "bhekul", "bhimal", "dhaman", and "pharan". This tree is useful in several ways. Its leaves are used for fodder, bark fiber after treatment is used for making rope, bags, nets, etc. and the sticks from which the bark has been removed are good as smokeless fuel. The tree has a great economic significance. Conventionally, the bast fibers from the stem are removed by dumping the stems in stagnant water bodies for months together. Following retting by the natural microflora present on the fibers, these fibers are separated by physical methods. The quality of the fibers obtained finally is not good due to prolonged incubation in the water body. Villagers sell these fibers and their products in rural markets.

2.6.2.2 Pretreatment of pectic wastewater

The wastewater from the citrus-processing industry contains pectinaceous materials that are barely decomposed by microbes during the activated-sludge treatment [224]. Tanabe et al. (1987) [225] has demonstrated a new wastewater treatment technique by utilizing an alkalophilic microorganism. Their soil isolates an alkalophilic *Bacillus* sp. (GIR 621), produced an extracellular endopectate lyase in alkaline media at pH 10.0. This treatment was successful for the eradication of pectic molecules from the wastewater.

2.7 Biotechnological degumming process

Alkaline pectinase is an emerging enzyme of commerce with primary employment in the textile and paper industries. Microorganisms are the main source of enzymes due to the usage of low-cost substrates [246]. The textile industry is the utmost polluting industry due to maximum chemical usage. Traditional scouring involves alkaline chemicals to expel non-cellulosic material for soft and hydrophilic fiber, suitable for industrial applications. Due to the usage of alkaline chemicals textile industries releases wastewater with high values of total dissolved solids, chemical oxygen demand, and biological oxygen demand [247-248]. Thus, the need of the hour for sustainable development is the replacement of chemical processes with eco-friendly biochemical processes. Bio scouring employs enzymes to remove impurities without disturbing fiber's structure, strength, and environment [249-250].

In the conventional degumming process, the remained gummy material is removed by chemical means. Onan industrial scale, the degumming of bast fibers is performed by treating the fibers with NaOH solutions (12- 20%) containing wetting and reducing agents [94]. After alkali treatment, these fibers are boiled, rinsed, neutralized, washed, and centrifuged several times. The fibers are then dried over a charcoal fire and treated and treated with softeners such as glycerin, and soap, etc. then cleaned fibers are further graded and processed.

Due to the consumption of energy and time, the biochemical activities of microorganisms for the degumming process are becoming more essential today. Response surface methodology (RSM) is a widely used statistical and mathematical technique for users to design experiments and optimize various process parameters. For dealing with multifactor experiments RSM, CCD was implemented to measure a second-degree polynomial model to optimize the response variables of the interest (reaction time and reaction temperature,etc), and finally, linear regression was used to find the results. RSM helps to decrease the number of experimental trials required to evaluate multiple parameters and their interactions. RSM helps to predict optimal value from the estimated surface shape if it is a simple hill. The estimated surface is complicated or away from the experiment region, the surface shape can be evaluated to confirm the direction in which new experiments should function

2.8 Industrial production and commercial suppliers of pectinases

Maximum commercially available pectinases are produced from the molds of the genus *Aspergillus*. Strains of *A. niger*, *A. oryzae*, *A. wentii* and *A. flavus* have been successfully utilized to produce a mixture of pectinases that have pH optima and other characteristics suited for fruit juice industries, which is their most common application. The commercial preparation available mostly contains a mixture of pectin esterase polygalacturonase and polygalacturonate lyase activity [251]. Industrially, the three different methods used to produce pectinases include surface bran culture (Koji method), deep tank (submerged method), and two-stage submerged process.

In the Koji method, all the ingredients are mixed in such a manner that the bran absorbs all liquid and in effect, the substrate becomes bran impregnated with various nutrients. Enzyme production takes one to six days and the process of separation from growth is cumbersome. On the other hand, in the submerged culture process, the cells are readily separated from the liquid growth medium. In the submerged culture process, carbohydrate sources such as glucose, molasses, com syrup, starch hydrolysates, milled cereal products, and wheat bran are used [252]. The nitrogen is derived from ammonium salts, com steep liquor, distillers soluble, yeast extract, gelatin, and casein, while mineral supplementation is in the form of inorganic salts. To produce pectinolytic enzymes, however, inducers such as beet, pulp, citrus peel, or dead sugar cossets and apple pomace must be added to the media.

The enzyme is sold either as a liquid preparation or as powder. Godfrey and West (1996) [253] have listed more than 15 companies worldwide offering pectinases in their sales programs. As many of these companies are agents and distributors. Table 2.3 whilst not warranted to be comprehensive, provides a guide to many sources of commercial pectinases.

Table 2.3: Commercial suppliers of pectinases.

Company	Country	Enzyme Name	Type and source	pH opt./range	Temp °C/range	Activity (suppliers) units	Applications
Advanced Biochemicals Ltd., Plot No. A-61, MIDC Area, Sinnar, Nasik.	India	Pectinase	-	-	-	-	Wine, juice.
Amano Pharmaceutical Co. 2-7, 1-Chome, Nishiki, Naka-Ku, Nayoga 460	Japan	Pectinase A	*Aspergillus* spp.	3.5	66	-	Juice, wine
		Pectinase P-11	*Aspergillus* spp.	4.5	55	-	Juice, wine
		Pectinase G	*Aspergillus* spp.	4.0	55	-	Juice, wine
		Pectinase GL	*Aspergillus* spp.	4.0	55	-	Juice, wine
		'Amino' enzyme PTE4	*Aspergillus* spp.	4.5	55	-	Juice, wine
		Pectinase P-11 'Amano'	*Aspergillus* spp.	3.5	55	-	Juice, wine
Biocatalysts Ltd., Main Avenue, Treforest Ind. Estate, Pontipridd, Mid Glamorgan CF37 5UT	UK	Macers FJ	Pectinase	4.8	60	60,000	Juiice, wine
		Pectinase 62L	*Aspergillus* spp.	4.8	60	20,000	
		Pectinase 162L	*Aspergillus* spp.	4.8	60	25,000	Juiice, wine
		Pectinase 444L	*Aspergillus* spp.	4.8	60	600	Juiice, wine

Company	Country	Product	Enzyme	pH	Temp	Dosage	Application
C.H. Boehringer Sohn, Ingelheim,	West Germany	Panzyme	-	-	-	-	Juice, wine -
Ciba-Geigy, A.G, Basal	Switzerland	Ultrazyme	-	-	-	-	-
Gist-Brocades NV, PO Box 1, NL-2600 MA, Deift	Netherlands	Klerzyme, Rapidase and Rapizyme Rapidase CPE Rapidase Apple Sauce	Pectinase *Aspergillus niger* Pectin esterase *Aspergillus niger* Pectin Methylesterase	4.0- 4.5 3.8-4.3 2.0-6.0	NS 10-60	NS NS	Juice, wine Juice, wine Specific for apple sauce production
Monoenzyme India Ltd. (MIL), Hydrabad.	India	Pectinase	NS	NS	NS	NS	NS
Novo-Nordisk AS, Novo Allee, DK-2880, Bags-vaerd	Denmark	Citrozyme Novo Clairzyme Pectinex Ultrazyme Vinozyme	pectinase *Aspergillus niger* pectinase *Aspergillus niger* peetinase *Aspergillus niger* peetinase *Aspergillus niger*	4.0-4.5 3.0-4.0 4.0-5.0	40-50 40 40-50	NS 5000 NS	Juice, wine Wine Wine

Company	Country	Product	Enzyme	pH	Temp	Activity	Application
		Vinoflow	peetinase *Aspergillus niger*	4.0-5.0	40	NS	
		Olivex	Peetinase *Aspergillus niger*				Wine
		Peelzyme	Pectinase/cellulase/ hemicellulase	3.0-4.0	30-40	NS	
		Pectinex AR	*Aspergillus aculeatus*	5.0-6.0	30-40	4000	Wine
		Pectinex Ultra SP-L	Pectinase/cellulase/ hemicellulase *Aspergillus niger*	3.0-4.5 4.5	NS	26000	Olive oil extraction. Citrus fruit peeling.
		Vinozyme EC	Pectinase/arbinase **00 z**	3.5-5.0 NS		NS	Apple and peer juicing. Fruit mashing Red grape Proicess ingq
			Pectinase/hemicellulase *Aspergillus niger*				
Rhone-Poulenc, ABMBrewing and Enzymes, Poleacre Lane, Woodley, Stockport, S K 6 1PO	UK	Peetinase CPT	*Penicillium* spp.	2.5-6.5	60	NS	Feed, juice, paper, wine.
Rohm GmbH, Postfach 4242, Kirshenallee,	West German	Rohapect Range Rohament	Peetinase *Aspergillus niger* Peetinase	4.0-4.5 4.0-4.5	40-45 50	NS NS	Food, fruit juice,

| Darmstadt | y | Max | | wine. Food, fruit juice, |

*NS (Not specified)

CHAPTER 3: RESEARCH METHODOLOGY

3.1 Sample collection

Many soil samples were selected and stored in the aseptic bag from the fruit and vegetable waste dump area in Solang Valley and Vashisht (Manali, Himachal Pradesh, India, Lattitude: 32.2396^0 N and Longitude: 76.9787^0 E).

3.2 Enrichment of alkaline pectinase producing microorganisms

The soil (5.0 g) was dissolved in 50 ml of deionized water and vortexed thoroughly. Then the sample was centrifuged for 5 min. at 1000 rpm. to settle down the soil particles. The supernatant was saved and centrifuged for 15 min. at 10,000 rpm. to settle down microbes. After that supernatant was worn-out, and the pellet was dissolved in 2.0 ml of 0.1 M Tris-HCl buffer (pH 9.0). Enrichment of pectinase-producing microorganisms from this suspension was carried out by inoculating 100 ml of 0.1 M tris buffer composed of 1.0% (w/v) pectin with 1.0 % (v/v) soil extract. After that, the samples were incubated for 48 h at 37°C with 100 rpm prolonged shaking.

3.3 Screening of alkaline pectinase producing bacteria

3.3.1 Composition of Yeast Extract pectin (YEP) medium

Pectin 0.25%

Yeast extract 1.0%

The final pH of the medium was adjusted to 9.0 with 0.1 N NaOH. YEP, the medium was solidified by the addition of agar (1.5%, w/v), if necessary.

3.3.2 Sample processing

The colonies on the serial dilution plates of 10^{-6}, 10^{-7}, 10^{-8}, and 10^{-9} were transferred onto another pectin agar plate and again incubated for 24 hours at 37° C. A clear zone surrounding the colony with the potassium-iodide solution confirmed enzyme activity (E. Merck (India) Ltd., Mumbai-400018, India). Soares *et al*, [254] illustrated pectin agar medium was selected for pure bacterial isolation and contains citrus pectin (1.0%), ammonium sulfate (0.14%), di-potassium hydrogen phosphate (0.6%), magnesium sulfate (0.01%), potassium dihydrogen phosphate (0.20%), agar-agar (2.0%) and pH set were 9.0.

The isolated colonies were selected from dilutions 10^{-6}, 10^{-7}, 10^{-8}, and 10^{-9} to procure perfect and clear bacterial cultures. The obtained single and clear colonies were subcultured in pectin agar medium for further studies and again tested for their potential to use pectin as a growth substrate and cultures maintained in nutrient agar slants.

3.4 Screening to produce other hydrolytic enzymes by alkaline pectinase producing isolates

3.4.1 Amylase

Microbial isolates were stabbed on nutrient agar plates with starch (1.0% w/v) then incubated for48 hours at 37°C. After 48 hours the iodine solution (3.0%, w/v iodine in 0.3% KI solution was flooded on the plates. Positive amylase producers were observed as a clear zone surrounding the colonies in contrast to a blue culture.

3.4.2 Cellulase

Nutrient agar plates supplemented with 1.0% carboxymethyl cellulose were streaked with the bacterial isolates and after 24 h incubation at 37°C the media plates were overflowed with congo red solution (0.5% w/v). After 20 minutes plates were washed with 1.0 N NaCl to remove unbound dye. A yellowish zone over the bacterial colonies against a red background confirmed the presence of cellulose.

3.4.3 Lipase

Lipase production was confirmed by observing a clear halo zone over the colonies on the nutrient agar plates supplemented with 1.0% emulsified tributyrin.

3.4.4 Protease

Milk agar plates (NA with skimmed milk, 5.0%, v/v) were stabbed with the bacterial isolates, andsucceeding 24 h incubationat37°C clear zones around the colonies confirmed the production of protease.

3.4.5 Tannase

Nutrient agar plates composed of 0.1% (w/v) tannic acid were stabbed with the bacterial isolates and after 48 h incubation at 37°C formations of halo zone over the colonies confirmed the tannase production.

3.4.6 Xylanase

Nutrient agar plates composed of 0.5% (w/v) xylan were stabbed with the bacterialisolates and after 48 h incubation at 37°C xylanase production was observed by staining plates with congo red.

3.5 Pectinase assay

Pectinase activity was estimated by the dinitrosalicylic acid (DNSA) method of Miller (1959) [255], modified by Aguilar and Huitron (1990) [90].

3.5.1 Preparation of DNSA (Dinitrosalicylic acid) reagent

Composition of dinitrosalicylic acid (per liter) is given below:

Sodium hydroxide	2.0gm
Sodium potassium tartarate	40gm
Sodium sulphite	0.1gm
Phenol	0.4gm
DNSA	2.0gm

Sodium hydroxide, sodium sulfite, and phenol were diffused in 300 ml distilled water in an Erlenmeyer flask and sodium-potassium tartarate was added slowly to this solution with constant stirring on a magnetic stirrer until dissolved. This flask was covered with black carbon paper for protection from light and then DNSA was added to it. The reagent so prepared was filtered to remove any undissolved material and subsequently saved in a brown bottle at 4°C for another use.

3.5.2 Other reagents preparation

Tris-HCl buffer (0.01 M, pH 9.0)

Stock A: 0.1 M Tris base i.e., 12.1 g Tris in 1 liter of distilled water.

Stock B: 0.1 M HCl i.e., 8.67 ml of HCl in 1 liter of distilled water.

Buffer preparation: 50 ml of A+ 29.2 ml of B were mixed and the final volume was adjusted to 200 ml with distilled water.

Polygalacturonic acid (PGA)

PGA (1.0 g, w/v) was dissolved in 100 ml of Tris-HCl buffer (pH 9.0).

Galacturonic acid (GA) stock solution

A stock solution (0.2 mg/ml) of galacturonic acid was prepared by dissolving 25 mg of GA in 100 ml of distilled water.

3.5.3 Assay procedure

An appropriately 100 pi diluted cell-free supernatant was incubated with 1.0% w/v PGA (100pi) for 5 minutes at 60°C under static conditions. After the addition of 500pi of DNSA, boiled for 15 minutes. This mixture was finally blended with deionized water to make a final volume of 5.0 ml. Then absorbency of the brownish color wasestimated at 530 nm by spectrophotometer (Shimazu UV-160 spectrophotometer). One unit of the enzyme was defined as the amount of enzyme, that catalyzed the production of 1.0 p mol of galacturonic acid per min at fixed pH.

3.6 Total protein estimation in cell free supernatant

The enzyme protein content was estimated by the Lowry *et al.* (1951) [256] method, by utilizing bovine serum albumin (BSA) (1.0 mg/ml) as standard.

3.6.1 Reagents

Reagent A: $CuSO_4.5H_2O$

1.0g

Deionised water

100ml

Reagent B: Potassium sodium tartarate 2-0g

Deionised water 100 ml

Reagent C: In 0.1N NaOH solution 2.0% (w/v) anhydrous sodium carbonate

Reagent D: Folin-Ciocalteu's reagent and deionized water in the ratio of 1:1

Working solution: Reagents A, B and C were mixed in the 1:1:98 ratio.

Bovine serum albumin standard: 1.0 mg/ml BSA standard was prepared in deionized distilled water.

3.6.2 Protein estimation procedure: Folin-Ciocalteu's reagent method

A suitably diluted test sample (1.0 ml) was incubated with 5.0 ml of working solution for 10 minutes at room temperature, reagent D (0.5 ml) was supplemented to this mixture and the contents were blended properly by vertexing followed by incubation for 30 minutes at 37°C. Finally, the optical density of the color developed was estimated at 660nm in a spectrophotometer (Shimazu UV-160 spectrophotometer) contrary to a blank treated in the same way but having deionized distilled water rather than protein.

3.7 Selection of promising isolates for detailed study

Bacterial isolates showing a promising zone of clearance on YEP medium plates were further studied concerning intracellular/extracellular nature of pectinase, and optimization of the physical parameters such as pH, and temperature using crude enzyme source.

3.7.1 Nature of pectinase enzyme

3.7.1.1 Extracellular pectinase

Selected isolates were grown in 50 ml of autoclaved (15 psi; 30 min) YEP medium contained in a 250 ml Erlenmeyer flask. 1% (v/v) overnight culture was used as standard inoculums. After 24 h of growth at 37°C, the culture broth was centrifuged for 10 minutes at 8000 rpm. And this cell-free supernatant was considered as crude pectinase formulation and pectinase activity was determined in this supernatant.

3.7.1.2 Cell bound pectinase

For the detection of cell-bound pectinase, 50 ml of autoclaved (15 psi; 30 min) YEP medium contained in Erlenmeyer flask (250 ml) was inoculated containing 1% (v/v) overnight grown culture media. After 24 h of growth at 37°C, the culture broth was centrifuged for 10 minutes at 8000 rpm. Then cell biomass (obtained from 1.0 ml culture) was rinsed twofold with 0.01 M Tris-HCl buffer with pH 9.0. This same buffer was utilized to suspend washed cells. Then pectinase activity was estimated in the cells by incubating 100 pi of the cell suspension directly with an equal volume of substrate (1.0% PGA).

3.7.1.3 Intracellular pectinase activity

Selected isolates were grown in 50 ml of autoclaved (15 psi; 30 min) YEP medium contained in a 250 ml Erlenmeyer flask. 1% (v/v) overnight culture was used as standard inoculum. After 24 h of growth at 37°C, culture media was centrifuged for 10 minutes at 8000 rpm. Then cell biomass (procured from 10.0 ml culture) was washed twice with 0.01 M Tris-HCl buffer set with pH 9.0, also suspended in the same buffer (10.0 ml). Sonication was utilized to lyse the cells, sonicator probe at a frequency of 20 kHz and vibration for 2X10 second was utilized and cell debris was removed by centrifugation for 15 minutes at 1800 rpm. Then the pectinase activity was checked in the supernatant.

3.7.2 Optimum pH for activity of crude pectinase enzyme

Crude pectinase enzyme was prepared by growing selected isolates 50 ml of autoclaved (15 lbs; 30 min) YEP medium contained in a 250 ml Erlenmeyer flask. 1% (v/v) overnight culture was used as standard inoculum. After 24 h of growth, culture was centrifuged for 10 minutes at 8000 rpm. Pectinase activity in the cell-free upper liquid part was assayed at different pH values (pH 5.0-11.0) by the preparation ofa 0.01 M of various buffers, like citrate phosphate buffer (pH 5.0), phosphate buffer (pH 6.0-7.9), Tris-HCl buffer (pH 7.5-8.5) and glycine NaOH buffer (9.0-11.0). Anappropriately diluted enzyme (100 pi) was then mixed with PGA (1%, w/v) and the assay was carried out at 40°Cfor the estimation of optimum activity pH.

3.7.3 Optimum temperature for the crude pectinase activity

In the cell-free supernatant (formulated as illustrated under section 3.7.2) pectinase activity was assayed at different temperatures from 30-90°C at optimized pH 9.0.

3.8 Identification and taxonomical classification of isolate P3

The morphology and general physiological features of the isolate P3 were carried out according to Bergey's Manual of Systematic Bacteriology [257]. The classic description of the identification order was as given below:

3.8.1 Growth parameters

25 ml of YEP broth (pH 9.0) in 100 ml Erlenmeyer flask was inoculated with the actively growing culture of isolate P3 and incubated at 37°C under stationary conditions. Given observation was considered for seven days:

- Development of pellicle at the surface
- Evolution of any characteristic smell.
- Cells are capsulated or non-capsulated.

3.8.2 Physiological and biochemical parameters

Characterization of the physiological aspects of Isolate P3 were considered concerning Gram staining, shape, size, cell diameter of the bacteria, and presence of spores and motility. Biochemical aspects were considered as given below:

3.8.2.1 Production of acid/gas from carbohydrates

Medium

Diammonium hydrogenphosphate	1.0 g
Magnesium sulphate	0.2 g
Potassium chloride	0.2 g
Yeast Extract	0.2 g
Agar	15 g
pH	9.0
Distilled water	1000 ml

15 ml of 0.04% (w/v) bromocresol purple.

The bacterial isolate P3 was inoculated in this medium. Durham tube were placed in the test tube for acid production.

3.8.2.2 Medium for utilization of organic compounds by bacterial isolate P3

Medium

Magnesium sulphate heptahydrate	0.2 g
Sodium chloride	1.0 g
Diammonium hydrogen phosphate	1.0 g
Potassium dihydrogen phosphate	0.5 g
Yeast Extract	2.0 g
pH	9.0
Deionized water	1000 ml

This medium was dispensed in sugar tubes, sterilized at 10 lbs for 30 min, and inoculated with the isolate P3.

3.8.2.3 Anaerobic growth media

Anaerobic growth of the isolate was checked in anaerobic agar medium containing:

Trypticase	20 g
Glucose	10 g
NaCl	5g
Agar	20 g
Sodium thioglycolate	2.0 g
Sodium formaldehyde sulfoxylate	1.0 g
pH	9.0
Distilled water	1000 ml

This medium was dispensed in 30 ml (15 cm x 1.7cm) tubes in amounts sufficient to give a 75 mm depth of medium and sterilized at 10 psi. for 30 min. Tubes were inoculated with a loopful of culture by stabbing to the base of the culture tubes.

3.8.2.4 Effect of pH on the growth of P3 isolate

Isolate P3 was inoculated in 50 ml nutrient broth contained in Erlenmeyer flasks (250 ml). After that, the pH of the medium was adjusted to various pHs (4.0-10.0) by 0.1N NaOH and 0.1N HCl, and growth was observed up to 72 h.

3.8.2.5 Effect of temperature on the growth of P3 isolate

Isolate P3 was incubated in the nutrient broth (pH 9.0) and placed at different incubation temperatures between 25°C-50°C for 24 h at 150 rpm.

3.8.2.6 Catalase production by P3 isolate

An isolated colony from 24 h old culture of isolate P3 grown on nutrient agar plates was mixed with a drop of 10% (v/v) solution of hydrogen peroxide on a glass slide. Production of gas bubbles indicated the production of catalase.

3.8.2.7 Oxidase test

An isolated colony from 24 h old culture of isolate P3 was applied to an oxidase strip dipped in a 1.0% aqueous solution of tetramethyl-p-phenylenediamine dihydrochloride. Development of purple color indicated positive test.

3.8.2.8 Nitrate reduction

Nitrite-free nitrate broth: 1.0 g of potassium nitrate per liter of nutrient broth.

Reagent A

Sulphonilic acid	0.8%
Acetic acid	5N

Reagent B

0.6% Dimethyl-a-naphthyl amine in 5N acetic acid.

Isolate P3 was inoculated into the nitrite-free nitrate broth then incubated for 24 h at 37°C. 1.0 ml of reagent A was added, and the contents were mixed. Thereafter, 1.0 ml of reagent B was mixed. The appearance of red color indicated the presence of nitrite, which has been formed by the reduction of nitrate. To the tubes not showing red color, 5 mg/ml powder zinc was supplemented, then tubes were kept aside for further observation. The emergence of red color indicated the nitrate existence in the medium and the absence of red color indicated a reduction of nitrate to nitrite.

3.8.2.9 Indole production

Isolate P3 was inoculated in an indole production medium (1.0% peptone water) and kept for 24 h at 37°C. Then the presence of indole was detected by mixing 2.0 ml Kovac's reagent into the broth, which produces pink color if indole is produced.

3.8.2.10 MR (Methyl Red) and VP (Voges-Proskauer) test

Isolate P3 was grown in 5 ml of VP broth containing (g/l): peptone, 7.0; glucose, 5.0; sodium chloride, 5.0; pH, 6.5. For the MR test 5 drops of methyl red solution (methyl red, 0. lg; ethanol, 300ml; deionized water, 200 ml) were added to VP broth. The appearance of red color confirms the positive test.

For the VP test, 3.0 ml of 40% NaOH and 1.0 mg creatine was added to the culture broth. The tube was placed for 30 min at room temperature and observed for the red color appearance.

3.8.2.11 Hydrolysis of urea

Isolate P3 was streaked onto Christensensen's medium slants that contained (g/1): peptone, 1.0; sodium chloride, 5.0; dipotassium hydrogen phosphate, 2.0; phenol red (1:500 in water), 6.0 ml; agar, 20; glucose, (10% solution) 10 ml; urea (20% solution), 100ml; pH, 6.9. Glucose and urea were sterilized separately and mixed to the basal medium. The appearance of purple, pink color, around and beneath the growth of isolate indicated hydrolysis of urea.

3.8.3 Molecular Characterization of alkaline pectinase producing P3 bacteria isolate

For the molecular characterization of the selected strain, 16S rDNA sequence analysis was executed. Chen and Kuo [258] method with small changes was used for obtaining genomic DNA. 1.0% agarose gel electrophoresis was performed for confirmation of the extracted DNA. Then after visualizing under UV light, PCR amplification was performed using the universal primer (delta company) along with 16S rDNA sequence analysis [259]. 50 microliters of the reaction mixture were prepared to contain dNTPs mix, 2.5 microliters genomic DNA template, 1.0 micro liter of 10 pM primers, and 2.5 U of DNA polymerase. Gene Amp PCR System 2700 Applied Biosystems was used for the amplification of the genomic DNA. 1.0% agarose gel electrophoresis was used for the PCR amplified production analysis. After visualizing under UV light, the purification of the PCR amplified product was performed according to the instructional protocol by manufacturer Promega. Amplified DNA fragment sequence analysis performed using 3130 Genetic Analyzer by Applied Biosystem DNA sequencer. MEGA 7 software was used for the phylogenetic and molecular evolutionary genetic analysis [260].

3.9 Effect of environmental parameters on pectinase production by *Bacillus tropicus* P3 submerged and solid-state fermentation conditions

3.9.1 Inoculum synthesis

For studying the effect of various environmental factors on pectinase synthesis by *Bacillus tropicus*, the inoculum was prepared by inoculating this bacterium into 50 ml of autoclaved (15 psi; 30 min) YEP medium contained in a 250 ml Erlenmeyer flask. 1% (v/v) culture media was used as standard inoculum whereas 50 ml YEP medium was used in all the experiments unless specified. The parameters studied were as follows:

3.9.2 Submerged fermentation (SmF)

The effect of environmental factors in SmF was investigated as follow:

3.9.2.1 Nutrient media

Various nutrient media supplemented with 0.25% pectin were tested to select a suitable media for pectinase production from *Bacillus* sp. P3. Different nutrient media tested were: Yeast Extract (YE), Nutrient Broth (NB), Malt extract, Luria Broth (LB), and Mueller Hinton Broth (MHB).

3.9.2.2 Effect of various concentrations of YE and pectin in YEP medium.

For the optimization of the optimum concentrations of yeast extract and pectin in the growth medium for maximal pectinase synthesis, different concentrations of yeast extract from 0.25-4.0%, w/v and pectin 0-0.5%, w/v were used. And further pectinase activity was estimated at 60°C and pH 9.0.

3.9.2.3 Effect of pH on pectinase production

To study the effect of the growth medium on enzyme production, the pH was fixed to various pH's (5.0-10.0) by 0.1N NaOH and 0.1N HCl. And further pectinase activity was estimated at 60°C and pH 9.0.

3.9.2.4 Effect of temperature on pectinase production

For the optimization of incubation temperature on enzyme production, the medium containing bacterial culture was incubated at different temperatures (25-50°C). Thereafter, pectinase activity was estimated at 60°C and pH 9.0.

3.9.2.5 Agitation and pectinase production

Agitation conditions favoring maximal production of pectinase were optimized in the YEP medium. This medium was inoculated with the standard inoculum of *Bacillus* sp. P3 (1%, v/v) and incubated (37°C) at different speeds (static-250 rpm). Thereafter, pectinase activity was estimated at 60°C and pH 9.0.

3.9.2.6 Inoculum size and pectinase production

For the optimization of the size of the inoculum 50 ml yeast extractpectin (YEP) medium was taken in

Erlenmeyer flasks (250 ml) with distinct inoculum sizes (0.5-5.0%, v/v). Then, pectinase activity was estimated.

3.9.2.7 Salts

The effect of various metal ions on pectinase production was observed by supplementing different metal ions (final concentration, 1mM) including $CaCl_2.2H_2O$, $MgSO_4.7H_2O$, $CuCl_2.2H_2O$, $CoCl_2.2H_2O$, $MnSO_2.4H_2O$, H_3BO_3, $ZnCl_2$, Fe (III) Citrate, $Na_2MO_4.2H_2O$, $FeSO_4$, KCl, and NaCl in yeast extract pectin medium followed by incubation at 37°C under shaking conditions (250 rpm) for 16 h. Optimized salts were incorporated into the YEP medium in different concentrations (0.05-3.0 mM).

3.9.2.8 Effect of Carbon sources on pectinase production

Various sugars (1.0% w/v, final concentration) including glucose, mannitol, glycerol, maltose, sucrose, lactose, cellobiose, fructose, xylose, starch, galactose, arabinose, rhamnose, galacturonic acid, and sodium acetate, were supplemented with 0.25% pectin under shaking conditions (150 rpm), for 16 hours at 37°C. After confirming best carbon sources, the best carbon sources were supplied in the YEP medium in concentrations ranging from 0.25-3.0% (w/v).

3.9.2.9 Effect of nitrogen sources and pectinase production

Different nitrogen sources such as casein, soybean meal, skim milk powder, peptone, tryptone, glycine, urea, ammonium chloride, ammonium nitrate, ammonium sulfate, and ammonium citrate were supplemented separately YEP medium for the optimization of the appropriate nitrogen source for the pectinase production.

3.9.2.10 Effect of above optimized components in various combinations on pectinase production

To study the effect of optimized components alone or in combinations on pectinase production, they were supplied in yeast extract pectin in optimized concentrations.

3.9.2.11 Growth curve and enzyme production by *Bacillus tropicus* P3 in optimized medium

The relationship of growth of *Bacillus* sp. P3 to pectinase production was studied in the optimized YEPC (YEP supplemented with 1 mM $CaCl_2.2H_2O$) medium at 37°C over 36h under shaking (250 rpm) conditions. Samples were withdrawn periodically to assess the final pH, biomass, pectinase activity, and protein content of the cell-free supernatant.

3.9.3 Solid state fermentation (SSF)

3.9.3.1 Enzyme production

Solid substrate (5.0 g) in Erlenmeyer flasks (250 ml) were moistened with deionized water and autoclaved for 30 minutes at 15 psi. 2.0 ml of the inoculum was added to each flask of solid substrate, thereafter, incubated at 37°C for a particularperiod. Samples were withdrawn at regular intervals to assess the pectinase production.

3.9.3.2 Extraction of enzyme

0.01M Tris-HCl buffer (25 ml) with pH 9.0 was added to all flasks and after thorough mixing, the contents were kept at 4°C in a refrigerator for half an hour under static conditions. The enzyme was separated by centrifugation of the suspension for 15 minutes at 8000 rpm. 25 ml of Tris-buffer was again added to the centrifuged solid moleculesfor the separation of the leftover enzyme if any. Pectinases so extracted were pooled and its final volume was noted down. Then pectinase activity was estimated by the Miller (1959) [255] method, modified by Aguilar and Huitron (1990) [90] with units U/g of solid substrate utilized.

3.9.3.3 Optimization studies for pectinase production by *Bacillus* sp. P3.

The effect of environmental factors in solid-state fermentation was investigated as follow:

3.9.3.3.1 Effect of different substrates and moisture contents

Solid substrate's effect on pectinase synthesis was observed by utilizing different substrates such as apple pomace, rice bran, and wheat bran. Pectinase synthesis at different moisture levels such as 50, 60, 65, 70, 75, and 80% of moisture content in standard-sized media by utilizing deionized water as moistening agent was observed utilizing wheat bran and rice bran.

3.9.3.3.2 Effect of salts

Different salts namely $CaCl_2.2H_2O$, $MgSO_4.7H_2O$, $COCl_2.2H_2O$, $MnSO_4.4H_2O$, H_3BO_3, $ZnCl$, KCl, and $NaCl$

(1mM) were suspended in deionized water in the solid substrate.

3.9.3.3.3 Carbon supplementation effect

In solid media various carbon sources including glucose, pectin, mannitol, galactose, polygalacturonic acid, maltose, and xylose were supplied individually to make the final 1% (w/v) concentration. Were suspended in the 15 ml deionized water per 5g of solid substrate for moisture adjustment.

3.9.3.3.4 Nitrogen supplementation effect

In solid media various carbon sources including yeast extract, peptone, tryptone, glycine, urea, ammonium sulfate, ammonium nitrate, and ammonium citrate were supplied individually to make the final 1% (w/v) concentration. Were suspended in the 15 ml deionized water per 5g of solid substrate for moisture adjustment.

3.10 Purification of pectinase of *Bacillus tropics* P3

3.10.1 Preparation of cell-free supernatant

For purifying the pectinase enzyme, *Bacillus tropics* P3 was grown in 250 ml yeast extract pectin medium at 200 rpm, 37°C for 18 h. The cell-free supernatant was procured after centrifugation (10,000 rpm, 15 min, and 4°C). This preparation was then utilized for enzyme activity and protein content by the methods described by the dinitro salicylic acid (DNSA) given by Miller [255] and by Lowry *et al* method (1951) [256] respectively.

3.10.2 Ammonium sulfate precipitation

Above mentioned cell-free supernatant was precipitated with $(NH_4)_2 SO_4$ (0-40 and 40-100% saturation) for the proteins present in it. Precipitation was carried out in a refrigerator with constant stirring and by adding small lots of ammonium sulphate. Thereafter mixtures were centrifuged at 11,000 rpm, 4°G for 20 minutes. Then protein precipitates collected from both cut-offs were suspended in a minimal amount of 0.01 M Tris-HCl buffer (pH 9.0).

3.10.3 Dialysis

Dialysis bags of convenient length were boiled in double distilled water with a pinch of $NaHCO_3$ and EDTA for 5 minutes. Thereafter properly washing with deionized water, the bags were tied at one end and the protein sample (obtained from ammonium sulfate precipitation) was added into the dialysis bags. These dialysis bags were kept in cold 500 ml of Tris-HCl buffer (0.01M, pH 9.0) and dialyzed under refrigeration conditions with continuous stirring. The buffer was changed every six hours and dialysis was carried out overnight. The completion of dialysis was confirmed by the addition of $BaCl_2$ (0.1M) to the dialyzing buffer. No precipitation of $BaSO_4$ in the buffer confirmed the removal of all sulfate ions. The dialyzed protein sample was analyzed for enzyme activity and protein content.

3.10.4 Concentration of dialyzed fraction

The fraction (40-100% ammonium sulfate saturated) showing alkaline pectinase activity was concentrated from 15.0 to 3.0 ml by ultrafiltration utilizingan Amicon unit along witha 10 kDa molecular weight check filter.

3.10.5 Purification

The concentrated protein sample was processed further for purification by ion exchange (DEAE Sephacel) and gel filtration (Sephadex G-100) chromatography

3.10.5.1   Ion exchangechromatography

By utilizing DEAE Sephacel, anion exchanger with diethyl amino ethyl as ion exchange group which remains charged throughout the working range, Ion exchange chromatography was performed.

DEAE is supplied pre-swollen in 24% ethanol. The slurry was prepared by pouring out the 24% ethanol solution further replacing it with Tris-HCl buffer (0.01M, pH 9.0) in a ratio of 25% buffer and 75% settled gel. Then slurry was washed thrice with this buffer to remove any residual ethanol. This slurry was packed into a glass column (15x0.55 cm, 10 ml bed volume). After that column was equilibrated with Tris-HCl buffer (0.01M, pH 9) until the pH of the effluent was the same as that of the in-going solution. 3.0 ml of the concentrated protein sample was loaded onto this column. The column was washed with three-bed volumes of equilibrating buffer to remove unbound proteins from the gel. The protein was then eluted using a 0 - 1.0 M NaCl linear gradient created in the equilibrating buffer. Various fractions of 2.5 ml volume were collected. Then absorbance of the individual fraction was taken at 280 nm on a spectrophotometer using equilibrating buffer as blank and then plotted against each fraction. The fractions of a single peak were pooled separately and estimated for pectinase activity and protein concentration by Lowry's (1951) [256] method. The fractions with pectinase activity were pooled, concentrated, and saved for further analysis.

3.10.5.2   Gel filtration chromatography

5.0 g of Sephadex G-100 powder was swollen in 0.01M Tris-HCl buffer (pH 9). The gel slurry was conserved for 5 h at 100°C in the water bath. After cooling the gel slurry to room temperature, it was packed in a glass column (35x1.5 cm, bed volume 60 ml) as per the instructions given in the purification protocol. The column was equilibrated with Tris-HCl buffer (0.01M, pH 9) to wash down any impurity. The concentrated protein sample as prepared in section 3.10.4 was loaded onto this column and the elution of the protein was done utilizingan equilibrating buffer movement of 20 ml h". Fractions of 2.5-ml volume were collected after the void volume (20 ml). Thereafter individual fraction absorbance was taken at 280 nm on a spectrophotometer using equilibrating buffer as blank and then plotted against each fraction. The fractions of a single peak were pooled separately and assayed for pectinase activity and protein content as described by the dinitro salicylic acid (DNSA) given by Miller [255] and Lowry *et al* method (1951) [256] respectively.

3.10.6 Sodium   dodecyl   sulfate-polyacrylamide   gel electrophoresis

The purified protein was loaded onto SDS-PAGE (10%) as described by Laemmli (1970) [261] to determine the protein profile by using electrophoresis apparatus as described below:

3.10.6.1 Reagents

A. Acrylamide-bisacrylamide solution

Acrylamide   30g

Bis-acrylamide   0.8g

Distilled water   100ml

This solution was stored, in a brown bottle for further use

B. Stacking gel buffer: 0.5 M Tris buffer.

Tris   6.05g

Distilled water   100ml

pH   6.8

C. Separating or running gel buffer: 1.5 M Tris buffer

Tris   18.15g

Distilled water   100m

pH   8.8

D. SDS: 10% (w/v) in deionised distilled water.

E. Ammonium per sulphate (APS): 10% (w/v) in deionised distilled water.

F. Tank buffer: A 4X stock solution of this buffer was prepared.

Tris   3.0g

Glycine   14.4g

Distilled water   250 ml

For use, 100 ml of this stock solution was mixed with 300 ml deionized water.

Finally, 4.0 ml SDS solution (10%) was mixed with the buffer.

G. Solubilizing buffer

0.062 M Tris	375mg
SDS	1.0 mg
Glycerol	5.0 ml
Bromophenol blue	0.01%
pH	6.8

H. Sample buffer

2-Mercaptoethanol	50pl
Solubilizing buffer	1.0 ml

3.10.6.2 Preparation of gel

3.10.6.2.1 Separating gel: For 10% mini gel, the separating gel solution was prepared by mixing:

Component volumes per gel mold		Volume of 10 ml (ml)
1.5 Tris-HCl buffer (pH 8.8)	:	2.5
Acrylamide-bisacrylamide solution	:	3.3
Double distilled water	:	4.0
SDS (10% w/v)	:	0.1
APS (10%)	:	0.1
TEMED	:	0.004

3.10.6.2.2 Stacking gel:

Component volumes per gel mold		Volume of 10 ml (ml)
Tri-HCl buffer	:	0.25
Acrylamide-bisacrylamide solution	:	0.33
Double distilled water	:	1.4
SDS (10% w/v)	:	0.02
APS (10%)	:	0.02

| TEMED | : | 0.0002 |

3.10.6.3 SDS-PAGE protocol

For casting and running SDS-PAGE Bangalore Ge Nei electrophoresis, mini gel apparatus was utilized. 0.75 mm thickness spacers were utilized to fastened the gel plates to the gel casting stand. 10% separating gel was poured between two plates and at room temperature allowed to polymerize for 30 minutes. Thereafter over the separating gel 5% stacking gel was poured and a comb was placed in the stacking gel then allowed to polymerize.

3.10.6.4 Sample preparation

The protein samples were suspended with sample buffer in a ratio of 1: 1, thereafter boiled for 4-5 minutes in the water bath and investigated on SDS-PAGE.

3.10.6.5 Electrophoresis

The sample (30 ml) was loaded onto SDS-PAGE in the individual well with a micro liter syringe. The whole gel unit was positioned in a buffer tank, containing running buffer then connected to the power supply. The gel was run at a constant current of 30 mA for stacking gel and 50 mA. The gel was gently withdrawn from the glass plate and finally stained by silver staining.

3.10.6.6 Coomassie blue staining

For staining the gel, it was fixed in fixing solution (methanol: acetic acid: water; 40: 15: 55) for 30 minutes the gel was rinsed thrice with double deionized water and placed in 200 ml of staining solution (0.1% Coomassie Brilliant Blue R-250, 50% methanol and 10% glacial acetic acid) for 20 minutes with gentle agitation. After that gel was destained in destaining solution (40% methanol and 10% glacial acetic acid) till the bands appear, replenished the solution several times until background of the gel is fully destained. The reaction was paused by 5% acetic acid solution.

3.11 Characterization of purified pectinase produced by *Bacillus tropicus* P3.

The purified alkaline pectinase enzyme was characterized concerning its optimum pH, temperatures, stability at various temperatures and pH values, as well as different metal ions, salts effect, and utilization of different substrates.

3.11.1 Effect of pH on pectinase activity

Alkaline pectinase activity was assayed at various pH values (pH 5.0-10.0) utilizing distinct buffers (0.01M) including citrate phosphate buffer (pH 5.0), phosphate buffer (pH 6.0-7.9), Tris-HCl buffer (pH 7.5-8.5), and glycine NaOH buffer (9.0-11.0). The assay was carried out at 45°C for the estimation of the optimum pH.

3.11.2 pH stability profile of pectinase

To determine pH stability, pectinase solutions prepared at various pH's from 4.0-12.0 were kept at room temperature for distinct time intervals up to 4 h. Thereafter, pectinase activity was assayed at 60°C.

3.11.3 Effect of temperature on pectinase activity

Alkaline pectinase activity from cell-free supernatant was estimated at various temperatures from 20-90°C at 9.0pH.

3.11.4 Thermostability profile of pectinase

To determine the thermostability of alkaline pectinase preparation was kept at different temperatures ranging from 45-90°C, for different time intervals up to 5 h. Thereafter, alkaline pectinase activity was assayed at 45°C and pH 9.0.

3.11.5 Effect of chelating agents and surfactants

For checking the effect of mercaptoethanol, urea, ascorbic acid, glycine, cysteine, EDTA, (final concentration, 1mM), and SDS (1%, w/v) tween's (20, 40, 60, and 80) were incorporated directly into the enzyme-substrate system.

3.11.6 Effect of metal ions on pectinase activity

Metal ions and salts at a final concentration of 1 mM in the assay mixture were studied separately for their effect on pectinase activity at 45°C and pH 9.0. Various metal ions and salts studied include Ca^{2+}, Mg^{2+}, Pb^{2+}, Co^{2+}, Cu^{2+}, Fe^{2+}, Cd^{2+}, Ni^{2+}, Ba^{2+} and Mn^{2+}.

3.11.7 Type of pectinolytic activity

To check the type of pectinase produced by *Bacillus* sp. P3, a modified method of Sherwood (1966) [262] was used. In brief, suitable diluted purified enzyme (prepared as mentioned in section (3.2.10.5) (250 pi) and PGA (1%, w/v) (250 pi) solution was incubated at 45°C for 10 minutes. A 100 pi sample from this solution was added to

500 pi of 0.5N HCl. One ml of 0.01 M thiobarbituric acid suspended indeionized water was added to it and the mixture was heated in a boiling water bath for 30 min, cooled, restored, and centrifuged for 15 minutes at 10,000 rpm. The optical density of the supernatant was determined by utilizing a Shimazu UV-160 spectrophotometer over the range of 480-580nm. The presence of the peak at 510 nm is indicative of hydrolase activity and the peak at 550 nm is indicative of lyase enzyme [126].

3.11.8 Determination of Michaelis-Menten constant (Km) and Vmax values

The Michaelis-Menten constants (Km and Vmax) values of purified pectinase were calculated from Lineweaver-Burk's plot between substrate (PGA) concentration (1-10 mg/ml) andrate of reaction.

3.12 Application of alkaline pectinase produced by *Bacillus tropicus* P3 in Degumming of buel (*Grewia optiva*) bast fibers

3.12.1 Bacterial treatment

By inoculating autoclaved 1.0 g of decorticated *Grewia optiva* fibers with 2% (v/v) bacterial culture, 2×10^2 CPU/ml in the final mixture was obtained. For optimum degumming, different moisture contents (80, 85, 90, and 95%) were adjusted in the beul fibers. And samples were withdrawn periodically for up to 2.40 hrs to calculate the amount of galacturonic acid produced. The final pH and dry weight of the fibers were assessed, and galacturonic acid concentration was estimated by the DNS method. The treated *Grewia optiva* fibers were air-dried and all the experiments were repeated.

3.12.2 Optimization of pectinase enzyme concentration and reaction time for optimum degumming of *Grewia optiva* fibers

The Baract-Pereira method was applied for enzymatic and chemical treatment of *Grewia optiva* fibers [241]. The pectinase level optimization for fiber treatment was implemented by treating 1 gm of *Grewia optiva* fibers with 10 ml of 0.01 M Tris-HCl buffer 9.0 pH with distinct levels of 100-500 U/ml. Thereafter reaction time, temperature, pH, and concentration of alkaline pectinase were varied according to the experimental design (Table 3.1). 0.1 ml sample fractions were taken periodically from the reaction mixture which was on the continuous stirring mode at 500 rpm. For qualitative and quantitative analysis supernatant was preserved after centrifugation at 3000 g for 5 minutes and the enzyme was separated.

3.12.3 Chemical treatment

Sharma [244] method was applied for the chemical treatment of the *Grewia optiva* fibers by incubating 2gm of decorticated beul with 10 ml NaOH (2% w/v) solution under static conditions for 96 h in 90° C. For assessing degumming samples were periodically withdrawn.

3.12.4 Experimental design

The experimental design techniques used for optimizing the process parameters for the maximum synthesis of galacturonic acid are studied with RSM and CCD. This system is optimizing the effective variables with a minimum number of experiments. The four independent variables are studied at five levels (−2, −1, 0, +1, +2) (Table 3.1). A five-level and four-factor central composite rotatable design (CCRD) requiring a total of 30 runs were used to determine the experimental data. Reaction pH (X_1), temperature (X_2, °C), incubation time (X_3, hr), and pectinase concentration (X_4, g/l) are selected as independent variables and yield of galacturonic acid (mmole/g) as dependent variables (Table 3.2).

Table 3.1. Experimental design of four process variables in terms of coded values

Variables Coded levels	pH (X_1)	Temperature (°C) (X_2)	Incubation Time (Hrs) (X_3)	Enzyme Concentration (g/l) (X_4)
-2	7	27	1	0.5
-1	8	32	1.25	1
0	9	37	1.50	1.5
+1	10	42	2.15	2
+2	11	47	2.40	2.5

Table 3.2. Experimental design matrix and the experimental and predicted CCD design responses.

Run	pH (X_1)	Temperature (°C) (X_2)	Incubation Time (hr) (X_3)	Enzyme Concentration (g/l) (X_4)	Yield (mmole/g) Experimental	Yield (mmole/g) Predicted
1	8	32	1.25	2.0	460	461
2	8	32	2.15	1.0	311	310
3	8	42	1.25	1.0	352	353
4	8	42	2.15	2.0	458	459
5	10	32	1.25	1.0	300	296
6	10	32	2.15	2.0	412	411
7	10	42	1.25	2.0	416	414
8	10	42	2.15	1.0	303	302

9	9	37	1.5	1.5	450	448
10	9	37	1.5	1.5	445	448
11	8	32	1.25	1.0	350	354
12	8	32	2.15	2.0	423	421
13	8	42	1.25	2.0	485	487
14	8	42	2.15	1.0	319	320
15	10	32	1.25	2.0	402	399
16	10	32	2.15	1.0	302	303
17	10	42	1.25	1.0	285	283
18	10	42	2.15	2.0	440	437
19	9	37	1.5	1.5	449	448
20	9	37	1.5	1.5	447	448
21	7	37	1.5	1.5	430	427
22	11	37	1.5	1.5	320	325
23	9	27	1.5	1.5	412	412
24	9	47	1.5	1.5	433	432
25	9	37	0.6	1.5	391	391
26	9	37	2.4	1.5	413	412
27	9	37	1.5	0.5	184	183
28	9	37	1.5	2.5	422	423
29	9	37	1.5	1.5	448	448
30	9	37	1.5	1.5	447	448

3.12.5 Statistical Analysis

The effects of process parameters on response were analyzed by the RSM procedure to fit the second-order polynomial equation. Statistica (Stat Soft, Inc., USA) software was used to evaluate the experimental data. The basic form of the model equation is

$$Y = \beta_0 + \sum_{i=1}^{4} \beta_i X_i + \sum_{i=1}^{4} \beta_{ii} X_i^2 + \sum_{i=1}^{4} \sum_{j=i+1}^{4} \beta_{ij} X_i X_j \qquad (1)$$

Where Y represents the predicted response; β_o is the mean effect, β_i, β_{ii}, β_{ij} are the regression coefficients for linear, quadratic, and cross-product coefficients and X_i and X_j represent the coded independent variables.

3.12.6 Analytical estimation of degumming of fibers

The assessment of the degumming was done after considering the amount of galacturonic acid produced due to hydrolysis of pectic substances and finding the reduced weight of the fibers. For achieving these samples were withdrawn up to 24 hrs from the bacterial and chemical treatment process.

CHAPTER 4: RESULTS

4.1 Isolation and screening of alkaline pectinase producing bacterial strain

Hundred different bacteria were isolated from the soil, cauliflower waste, apple waste and citrus fruit waste and primarily screened based on pectinolytic activity by using the spread plate method. Alkaline pectinase-producing bacteria were isolated by plating the samples on a YEP medium and incubating at 37°C, pH 9. Only thirty isolates showed a clear zone around the bacterial colonies, named P1 to P30, and confirming the capacity of these isolates to synthesize alkaline pectinase (Table 4.1).

Out of these 30 strains, 1- 3 was from cauliflower, 4 -7 were from apple waste, 8-12 were from citrus fruit and 12 -30 were from soil samples. Four isolates (P3, P16, P21, and P27) were found to possess pectinolytic activity. The bacterial strains showing pectinolytic activity were selected based on the largest hydrolysis zone around its colony in the potassium-iodide assay and evaluate to the other strains (Figure 4.1). Among them, the P3 isolate displayed the largest hydrolysis zone and thus selected for further pectinase production, and studies were continued with this strain. It was observed that isolates No P3, P16, and P27could produce lipase, amylase, and xylanase besides other hydrolytic enzymes (Table 4.2).

Table 4.1: Bacterial isolates with pectinolytic activity from domestic and natural sources

Bacterial Isolates	Domestic and natural sources	Zone of Hydrolysis (mm)
P1	Cauliflower waste	17
P2		18
P3		25
P4	Apple waste	16
P5		16
P6		15
P7		16
P8	Citrus fruit waste	14
P9		14
P10		18
P11		12
P12		12
P13	Soil sample from Solang Valley	17
P14		16

P15	14
P16	19
P17	12
P18	12
P19	10
P20	13
P21	14
P22	13
P23	14
P24	15
P25	16
P26	17
P27	20
P28	18
P29	17
P30	18

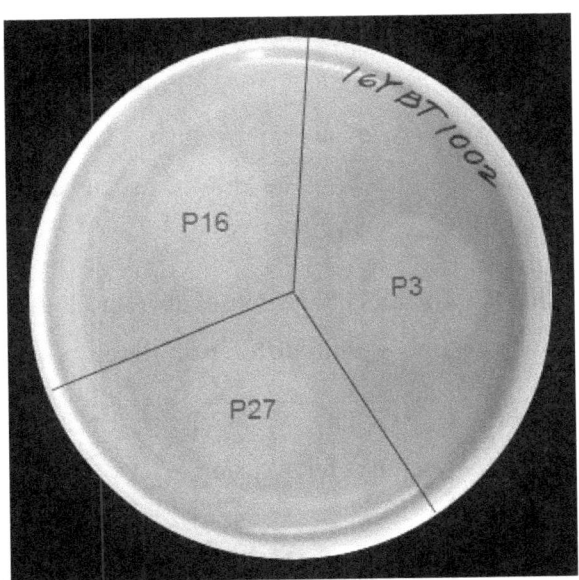

Figure 4.1: Pectin agar plate showing alkaline pectinase producing bacterial isolate stored after the qualitative screening

Table 4.2: Production of hydrolytic enzymes by selected isolates in a solid medium.

Isolate No.	Lipase	Amylase	Cellulase	Pectinase	Xylanase	Protease	Chitinase	Tannase
P3	+	+	+	+	+	+	-	-
P16	+	+	+	+	+	+	-	-
P27	-	+	+	+	-	+	-	-

=> The isolates were plated on NA supplemented with the individual substrate as outlined in Materials and Methods (section, 3.4).

=> Incubation conditions: 37°C, 48 h and pH 9.

4.2 Gram's character, shape, and localization of pectinase from different isolates

All the isolates were observed to be Gram-positive rods. Alkaline Pectinase production by all the isolates in the YEP medium was extracellular as no pectinase activity could be detected in intact cells and in the cell-free supernatant obtained after disruption of the cells (Table 4.3).

Table 4.3: Gram's character, shape, and localization of pectinase from different isolates.

Isolate No.	Gram character	Cell shape	Localization of pectinase
P3	+	Rods	Extracellular
P16	+	Rods	Extracellular
P27	+	Rods	Extracellular

4.3 Effects of pH and temperature on the pectinase activity of bacterial isolates

The increase or decrease in enzyme activity present in the cell-free supernatant of bacterial isolates was checked by regulating the reaction mixture pH between 5.0 - 10.0 and temperature between 30-90°C. Pectinase produced by most of the isolates was optimally active at pH above 8.0 (Table 4.4). Lowering the assay pH decreased the enzyme activity drastically. At pH 9.0, maximal pectinase production was shown by isolate P3 (11.0 U/ml) followed by isolate P27 (9.2 U/ml). Pectinase produced by most of the isolates except No. P10 and P23 were active up to 70°C (Table 4.5). Increasing the temperature above this temperature or decreasing it below 40°C decreased the enzyme activity drastically. Isolate P3, which was maximally active at 60°C (12.4 U/ml) showed activity up to 90°C for 5 min. As isolate P3 produced a large amount of pectinase which was alkalophilic (optimum pH, 9.0) as well as thermo tolerant (optimum temperature, 60°C), this bacterium was selected for further studies.

Table 4.4: Pectinase activities of cell-free supernatant from various isolates at different pH values.

Isolate No.	Pectinase activity (U/ml)						
	pH 5.0	pH 6.0	pH 7.0	pH 8.0	pH 9.0	pH 10.0	pH 11.0
P3	0.6	1.7	3.0	8.0	11.0	9.5	8.2
P16	-	1.9	4.2	8.2	8.5	2.4	2.1
P27	0.5	1.2	2.0	5.0	9.2	8.6	7.2

Table 4.5: Pectinase activities of cell-free supernatant from various isolates at different temperatures.

Isolate No.	Pectinase activity (U/ml) at various temperatures (°C)

	30	40	50	60	70	80	90
P3	6.5	10.5	11.2	12.4	7.9	5.4	2.0
P16	-	1.9	8.2	9.2	8.5	4.3	-
P27	2.7	6.5	8.6	9.5	7.6	5.0	1.5

4.4 Identification and taxonomical classification of isolate P3

4.4.1 Morphological and Biochemical Characteristics of the isolate

For the recognition of the selected isolate P3 the two conventional microbiological and current molecular technologies were used. The selected isolate based on the observed morphological and biochemical characterization (Table 4.6) compared with Bergey's [263] Manual of determinative bacteriology standard characterization, the isolate P-3 preliminarily identified as *Bacillus* sp. [264].

Table 4.6: Characteristics of the potent alkaline pectinase producing isolate

Colony Morphology	Opaque, dull, finally wrinkled, and adherent colonies
Cellular Characteristics	
Gram's staining	Positive
Motility	Motile
Morphology	Rods with rounded ends
Size	0.5-1.0 μm in length
Biochemical Reactions	
Amylase	+++++
Oxidase	+++++

Catalase	+++++
Indole production	--------
Voges-Proskauer	+++++
Citrate utilization	--------
Urea	--------
H_2S production	--------
Fermentation Reaction	
Glucose	Acid production
Galactose	Acid production
Maltose	Acid production

+++++ Positive; -------- Negative

4.4.2 Molecular Characterization of the isolate P3

Isolate's genomic DNA was employed as a template for 16S rDNA amplification by PCR and agarose gel electrophoresis was implemented for the examination of the PCR product (Fig. 4.2). The PCR product was first purified to remove PCR reaction components and sequenced on GATC company by ABI 3730XL DNA sequencer using forward and reverse primers. To GenBank database 16S rDNA nucleotide sequence was submitted with accession number MK332379 (Fig. 4.3).

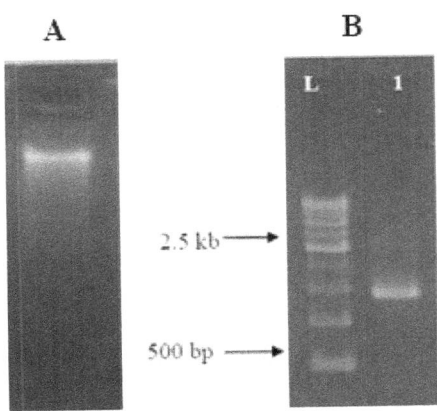

Figure 4.2: (A) DNA obtained from *Bacillus tropicus* strain MCCC 1A01406 (B) *Bacillus tropicus* 16S rDNA PCR product in agarose gel electrophoresis

```
  1 tgctcttatg agttagcggc ggagggtgag taacacgtgg gtaacctgcc cataagactg
 61 ggataactcc gggaaaccgg ggctaatacc ggataacatt ttgaaccgca tggttcgaaa
121 ttgaaaggcg gcttcggctg tcacttatgg atggacccgc gtcgcattag ctagttggtg
181 aggtaacggc tcaccaaggc aacgatgcgt agccgacctg agagggtgat cggccacact
241 gggactgaga cacggcccag actcctacgg gaggcagcag tagggaatct tccgcaatgg
301 acgaaagtct gacggagcaa cgccgcgtga gtgatgaagg ctttcgggtc gtaaaactct
361 gttgttaggg aagaacaagt gctagttgaa taagctggca ccttgacggt acctaaccag
421 aaagccacgg ctaactacgt gccagcagcc gcggtaatac gtaggtggca agcgttgtcc
481 ggaattattg ggcgtaaagc gcgcgcaggt ggtttcttaa gtctgatgtg aaagccccg
541 gctcaaccgt ggagggtcat tggaaactgg gggacttgag tgcagaagag gagagtggaa
601 ttccatgtgt agcggtgaaa tgcgtagaga tgtggaggaa caccagtggc gaaggcgact
661 ttctggtctg taactgacgc tgaggcgcga aagcgtgggg agcgaacagg attagatacc
721 cttgggtagt ccacgccgta aacgatgagt gctaagtgtt agagggtttc cgccctttag
781 tgctgcagtt aacgcattaa gcactccgcc tggggagtac ggtcgcaaga ctgaaactca
841 aaggaattga cggggcccg cacaagcggt ggagcatgtg gtttaattcg aagcaacgcg
901 aagaaccttа ccaggtcttg acatcctctg acaacсctag agatagggct tccccttcgg
961 gggcagagtg acaggtggtg catggttgtc gtcagctcgt gtcgtgagat gttgggttaa
1021 gtcccgcaac gagcgcaacc cttgatctta gttgccagca ttcagttggg cactctaagg
1081 tgactgccgg tgacaaaccg gaggaaggtg gggatgacgt caaatcatca tgcccctат
1141 gacctgggct acacacgtgc tacaatgggc agaacaaagg gcagcgaagc cgcgaggcta
1201 agccaatccc acaaatctgt tctcagttcg gatcgcagtc tgcaactcga ctgcgtgaag
1261 ctggaatcgc tagtaatcgc
```

Figure 4.3: 16S rDNA nucleotide sequence of *Bacillus tropicus* 1A01406

In the NCBI database, BLAST displayed significant alignment of *Bacillus tropicus* with *Bacillus paramycoides* strain MCCC 1A04098 97% similarities. And finally, the phylogenetic tree that determined the isolate is *Bacillus tropicus* (Fig. 4.4) [265].

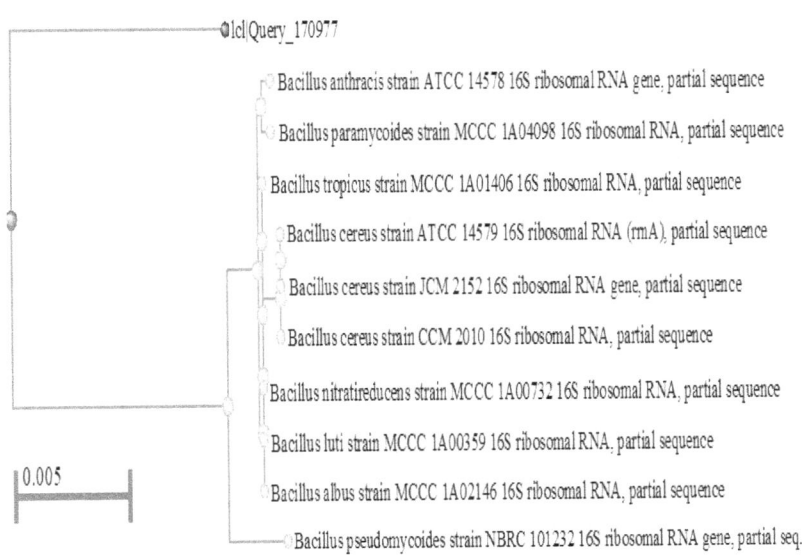

Figure 4.4: Phylogenetic tree exhibiting the lineage of *Bacillus tropicus* MCCC 1A01406 with another species

4.5 Effect of environmental parameters on pectinase production in submerged and solid-state fermentation systems

4.5.1 Submerged fermentation

4.5.1.1 Effect of growth media

Different media such as Malt extract, Nutrient Broth, yeast extract media, Luria-Bertani Broth were used to determine optimum media for alkaline pectinase production. Bacterial isolate was inoculated into 50 ml of these broths. The maximum growth was observed in Yeast Extract media supplemented with pectin proved to be the finest for the alkaline pectinase production under shaking condition about 21 U/ml of alkaline pectinase was produced while the alkaline pectinase production was minimum in Luria-Bertani Broth (Figure 4.5). The findings are also supported by the results of Oumer and Abate [266]; they have also received the highest pectinase production using yeast extract medium 10.1±1.44 U/ml [267].

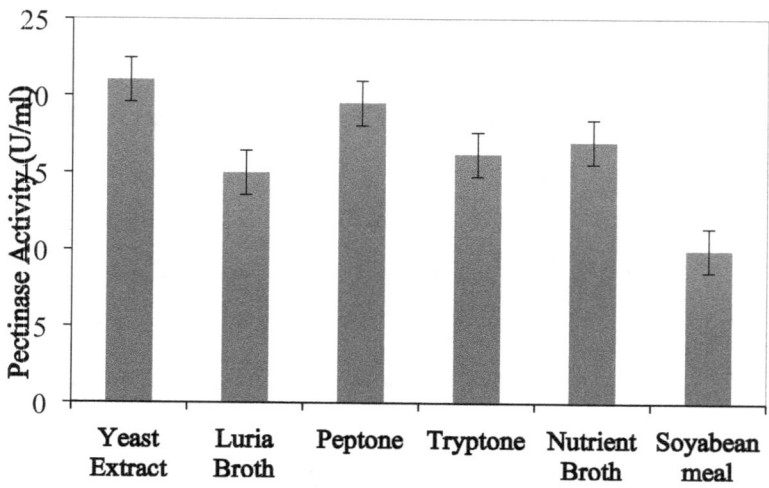

Figure 4.5: Effect of growth media on alkaline pectinase production by *Bacillus tropicus* MCCC 1A01406. Bars correspond to standard deviation.

For the optimization of the yeast extract and pectin concentration YEP medium for maximal pectinase production by *Bacillus tropicus* MCCC 1A01406, different concentrations of yeast extract (0.5-4.5%, w/v) and pectin (0-0.5%, w/v) utilized. The maximum pectinase activity (14.7 U/ml) was estimated in 1.5% (w/v) yeast extract and 0.25% pectin (Figure 4.6-4.7). Although increasing YE concentration (by 4%) increased biomass of the cells (OD_{600nm}, 1.02) but there was no increase in the pectinase production.

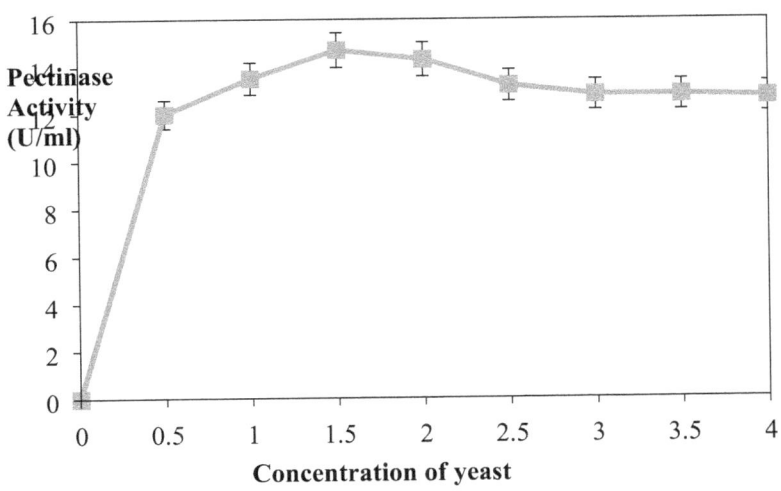

Figure 4.6: Effect of different concentrations of yeast extract on alkaline pectinase production by *Bacillus tropicus* MCCC 1A01406

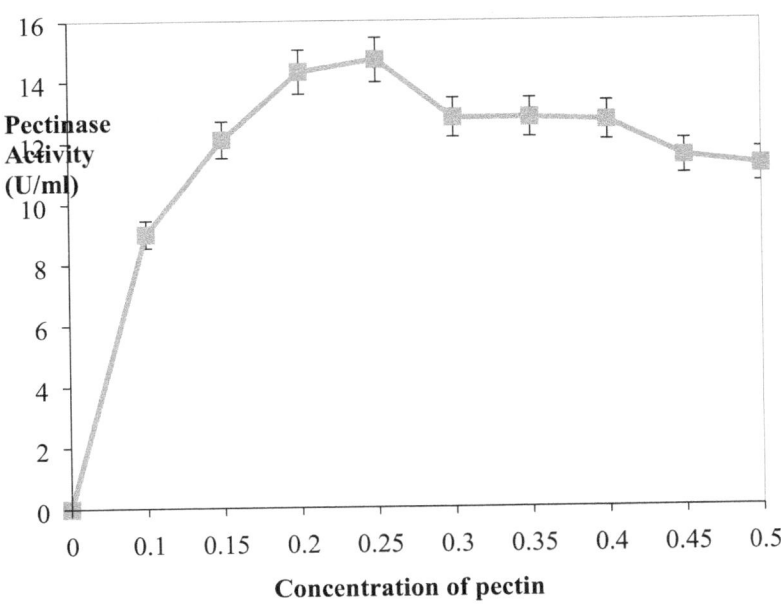

Figure 4.7: Effect of different concentrations of pectin on alkaline pectinase production by *Bacillus tropicus* MCCC 1A01406

4.5.1.2 Effect of media pH on pectinase activity

Different pH's effect on pectinase synthesis by *Bacillus* sp. P3 was studied by regulating yeast extract pectin medium between pH values 5.0-11.0. Maximum pectinase production (14.06 U/ml) was observed at pH 9.0, under shaking condition (150 rpm) after 24 h of incubation at 37°C, which coincided with maximum cell growth (OD_{600nm}, 0.7, Figure 4.8). Adjusting the pH of medium either towards acidic (5.0-6.0) or alkaline (9.0-10.0) side, resulting in decreased production of pectinase. A decrease of nearly 96% in pectinase production was observed when the pH of the YEP medium was modified either to 5.0 or 11.0.

4.5.1.3 Effect of Incubation temperature on pectinase activity

Incubation temperature effect on pectinase synthesis from *Bacillus* sp. P3 was observed from 25-50°C. It was identified that maximum growth (OD_{600nm}, 0.9) of *Bacillus* sp. P3 as well as maximal pectinase (15.2 U/ml) production took place at 37°C followed by 42°C (11.0 U/ml) and 32°C (9.5 U/ml) (Figure 4.9).

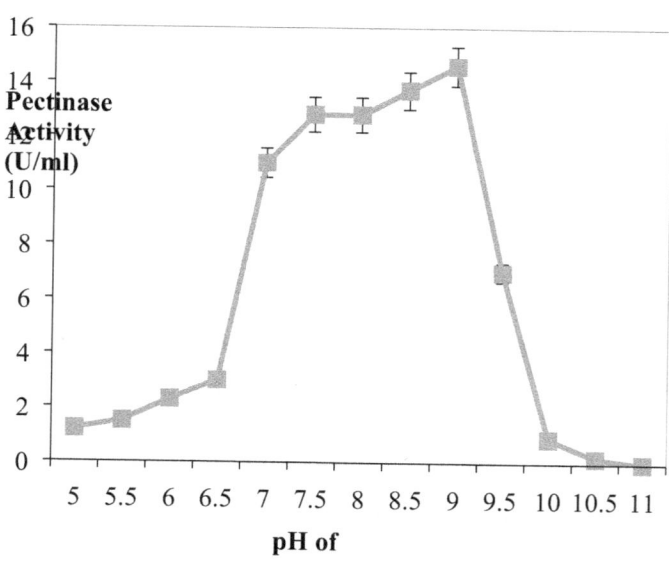

Figure 4.8: Initial pH of YEP medium effect on alkaline pectinase production by *Bacillus tropicus* MCCC 1A01406

Increasing the incubation temperature above 42°C decreased the pectinase production drastically and no activity could be detected in the culture incubated at 55°C.

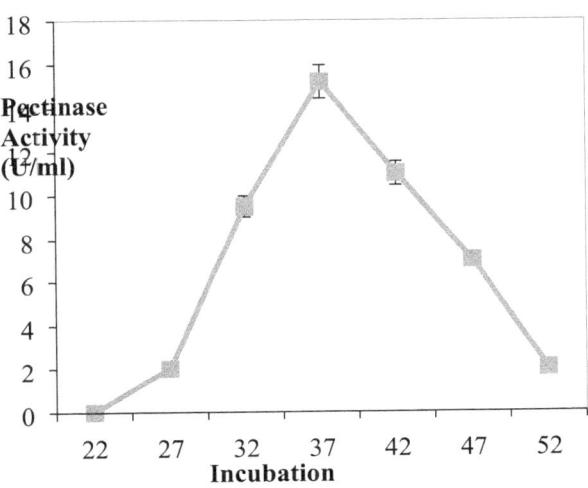

Figure 4.9: Incubation temperatures effect on alkaline pectinase synthesis by *Bacillus tropicus* MCCC 1A01406

4.5.1.3 Agitation

The agitation has a powerful significance on pectinase production. Although culture grown under static conditions produced pectinase (7.62 U/ml), agitation enhanced pectinase production by *Bacillus tropicus* MCCC 1A01406 considerably. Maximal enzyme production (16.2 U/ml) was observed at 250 rpm after 16 h incubation. Further more increase in the agitation decreased pectinase production, but a further decrease in production time was also observed (Figure 4.10).

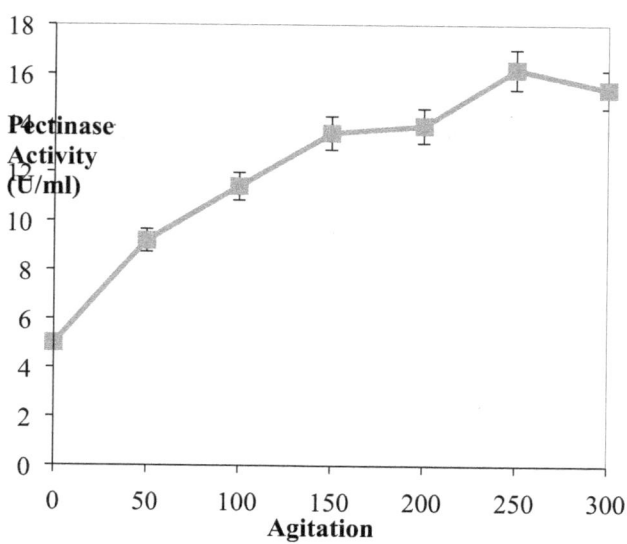

Figure 4.10: Agitation effect on alkaline pectinase synthesis by *Bacillus tropicus* MCCC 1A01406

4.5.1.4 Inoculum size

Various inoculum size's effect on pectinase synthesis was studied in YEP medium under shaking (250 rpm) conditions at 37°C up to 24 h of incubation. Maximal pectinase production (16.4 U/ml) was observed at 2% (v/v) inoculum size in 16 h of incubation followed by 2.5% (16.0 U/ml) and 3% (15.4 U/ml) at 16 h and 12 h of incubation respectively (Figure 4.11).

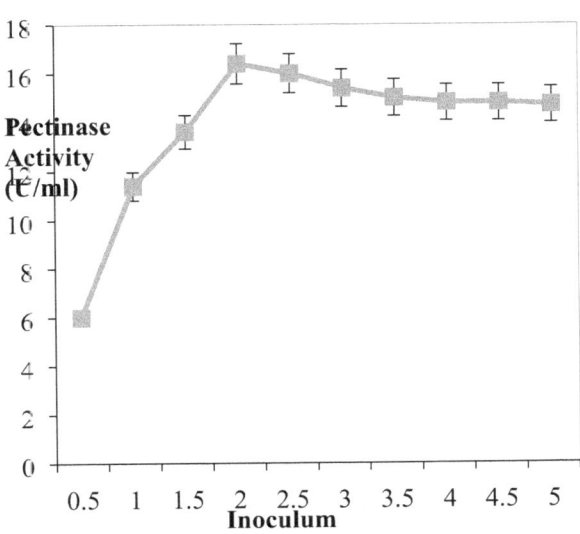

Figure 4.11: Inoculums size effect on alkaline pectinase production by *Bacillus tropicus* MCCC 1A01406 at 37°C, pH 9 under shaking (250 rpm).

4.5.1.5 Salts

Various salts effect on pectinase synthesis was studied in yeast extract pectin medium under shaking (250 rpm) conditions for 16 h. The addition of 1mM (final concentration) of $CaCl_2.2H_2O$ or $MgSO_4.7H_2O$ to the YEP medium resulted in enhanced pectinase production (Figure 4.12). To optimize the concentrations of salts for maximal pectinase production, different concentrations (0.5- 3.0 mM) were tested by supplementing them in a YEP medium. Maximum pectinase production (34.2 U/ml) was observed at the level of 1mM of $CaCl_2.2H_2O$ (Figure 4.13), whereas 1.0 mM $MgSO_4.7H_2O$ was observed to be best for maximal (31.6 U/ml) pectinase production (Figure 4.14).

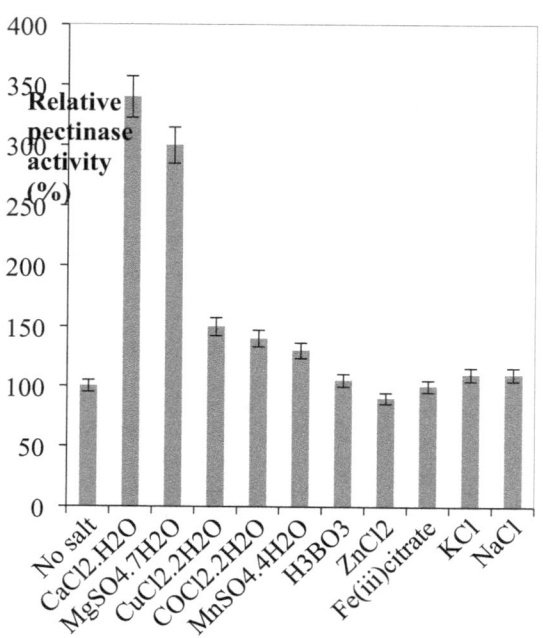

Figure 4.12: Production of alkaline pectinase from *Bacillus tropicus* MCCC 1A01406 in YEP medium using various salts (1mM) at 37°C, pH 9 under shaking conditions (250 rpm) for 18 h.

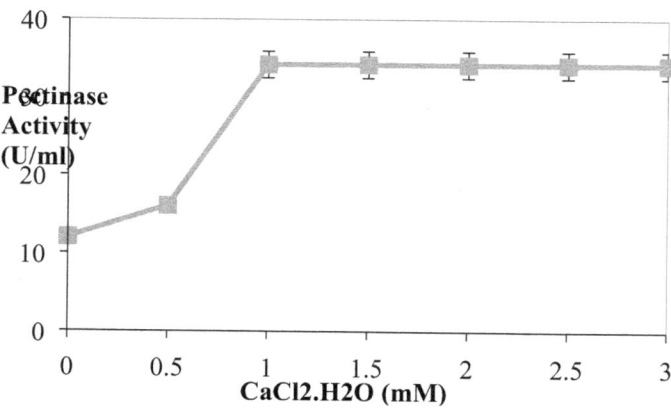

Figure 4.13: Effect of different concentrations of $CaCl_2.2H_2O$ on pectinase production by *Bacillus tropicus* MCCC 1A01406 in YEP medium at 37°C, pH 9 under shaking conditions (250 rpm)

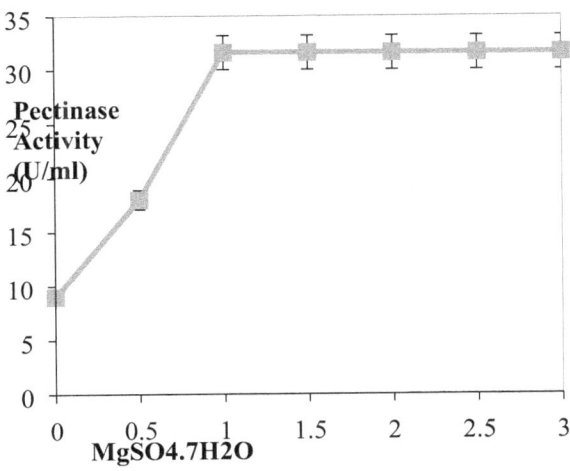

Figure 4.14: Effect of different concentrations of MgSO₄.7H₂O on pectinase production by *Bacillus tropicus* MCCC 1A01406 in YEP medium at 37°C, pH 9 under shaking conditions (250 rpm)

4.5.1.6 Carbon sources

The production of pectinase by *Bacillus* sp. P3 using various carbon sources (glucose, mannitol, glycerol, maltose, sucrose, lactose, cellobiose, fructose, xylose, starch, galactose, arabinose, rhamnose, galacturonic acid, and sodium acetate, final conc. 1%, w/v) was studied in YEP medium at 250 rpm, 37°C for 16 h. Out of various carbon, sources studied mannitol increased pectinase activity by 0.3 fold as compared to YEP medium. Most of the other carbon sources used except polygalacturonic acid (PGA) inhibited the enzyme production (Figure 4.15). In another experiment, it was found that 0.5% (w/v) mannitol in the yeast extract pectin medium yielded maximal pectinase production (30.5 U/ml) (Figure 4.16).

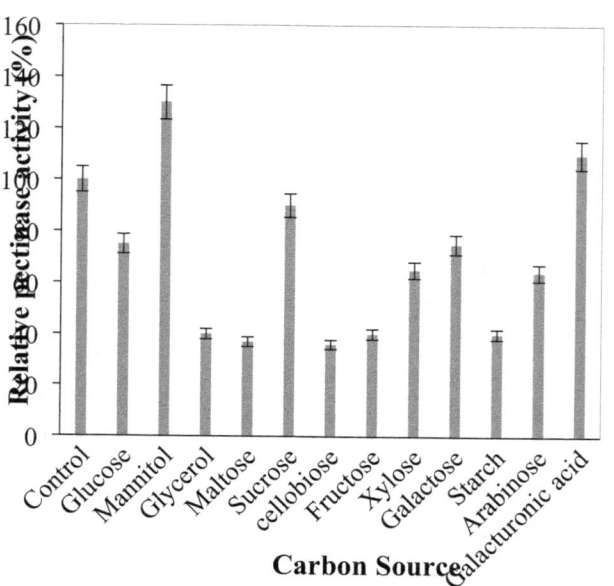

Figure 4.15: Pectinase production from *Bacillus tropicus* MCCC 1A01406 in YEP medium utilizing different carbon sources (1% w/v) at 37°C, pH 9 under shaking conditions (250 rpm) for 18 h

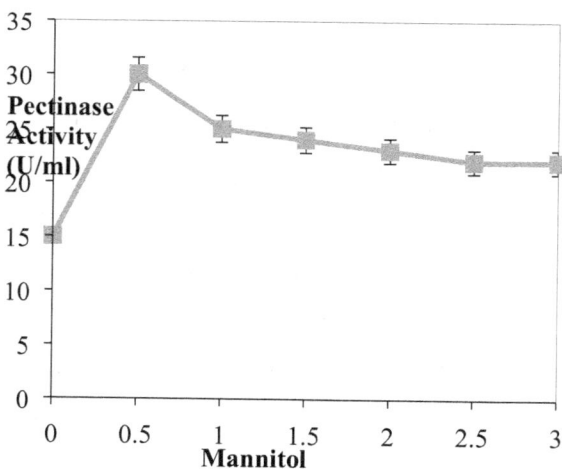

Figure 4.16: Effect of various concentrations of mannitol on alkaline pectinase production by *Bacillus tropicus* MCCC 1A01406 in YEP medium

4.5.1.7 Nitrogen sources

To select a suitable nitrogen source for maximal pectinase production, various organic and inorganic nitrogen sources (1%, w/v) were supplemented in YEP medium under shaking (250 rpm) conditions for 16 h at 37°C. Glycine, urea, and ammonium citrate inhibited cell growth as well as pectinase production. Whereas other nitrogen sources (peptone, tryptone, casein, soybean meal, ammonium nitrate, ammonium sulfate, and skim milk powder) decreased pectinase production up to 84% (Figure 4.17).

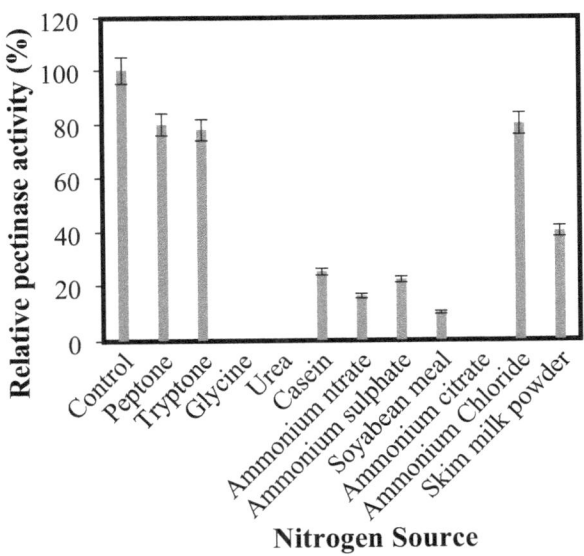

Figure 4.17: Production of alkaline pectinase from *Bacillus tropicus* MCCC 1A01406 in YEP medium using various nitrogen sources (1% w/v) at 37°C, pH 9 under shaking conditions (250 rpm) for 18 h

4.5.1.8 Effect of optimized components on alkaline pectinase synthesis

Optimized components effect alone or in combinations was studied in YEP medium with constant shaking (250 rpm) for 16 h at 37°C). Of the different components supplemented 1mM $CaCl_2.2H_2O$ produced maximal (55.7 U/ml) pectinase followed by combination of $MgSO_4.7H_2O$ + $CaCl_2.2H_2O$ + mannitol (55.8 U/ml) (Table 4.7).

Table 4.7: Optimized components effect on alkaline pectinase production by *Bacillus tropicus* MCCC 1A01406 in YEP medium using various nitrogen sources (1% w/v) at 37°C, pH 9 under shaking conditions (250 rpm) for 18 h

Components	Biomass O.D.$_{600nm}$	Final pH	Enzyme activity (U/ml)
Control	0.75	9.2	19.0
$CaCl_2.2H_2O$ (1.0 mM)	0.70	9.8	55.7
$MgSO_4.2H_2O$ (1.5 mM)	0.66	8.9	44.0
Mannitol (0.5%, w/v)	0.79	8.82	30.2
$CaCl_2.2H_2O$ + $MgSO_4.2H_2O$	0.72	8.9	54.8
$CaCl_2.2H_2O$ + Mannitol	0.75	8.93	53.4
$MgSO_4.2H_2O$ + Mannitol	0.74	9.1	45.6
$CaCl_2.2H_2O$ + $MgSO_4.2H_2O$ + Mannitol	0.79	9.3	55.8

4.8.1.9 Growth curve and pectinase production by by *Bacillus tropicus* MCCC 1A01406 in YEPC medium

The relationship of growth of *Bacillus* sp. P3 to pectinase production was studied in YEPC (YEP medium supplemented with 1mM $CaCl_2.2H_2O$) medium at 37°C over 36 h, under shaking conditions (250 rpm). The culture achieved maximum growth (1.1, OD_{600nm}) after 12 h of incubation. However, further, incubation caused a decline in the culture absorbance. Post 36 h incubation, the cell biomass was reduced to nearly half (0.52, $OD_{600}nm$) of the maximum value. Pectinase production by *Bacillus tropicus* MCCC 1A01406 was directly proportional to time up to 12 h of incubation. Although further incubation resulted in decreased biomass, pectinase activity kept on increasing, reaching a maximum of 65.0 U/ml after 20 h of incubation. This value of pectinase corresponded to the maximal level of protein content (2.7 mg/ml) in the culture supernatant. After 6 h of incubation, the pH value was 6.4. Subsequently, the pH value increased with an increase in the incubation period, reaching maximum (pH, 9.2) after 24 h of incubation (Figure 4.18).

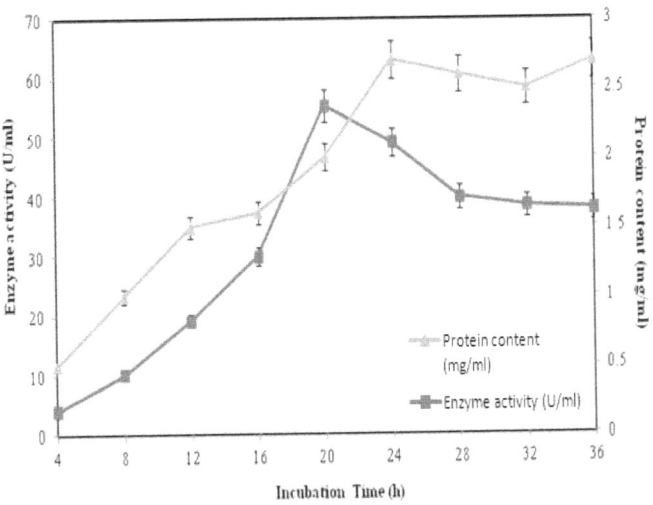

Figure 4.18: Growth curve and alkaline pectinase synthesis by by *Bacillus tropicus* MCCC 1A01406 in YEPC medium

4.5.2 Solid state fermentation (SSF)

4.5.2.1 Effect of different substrates and moisture content

The SSF was carried out using different substrates. Wheat bran as the prime solid substrate yielded 4600 U of pectinase g'1 dry substrate after 36 h of incubation at 37°C (75% moisture level) (Figure 4.19). And with rice bran and apple pomace maximal pectinase produced were 3265.25 and 78.0 U/g at 65 and 80 % moisture respectively after 48 h incubation (Figure 4.20-4.21). A decline in activity was observed after 65% moisture content in wheat bran and after 80% moisture content in rice bran.

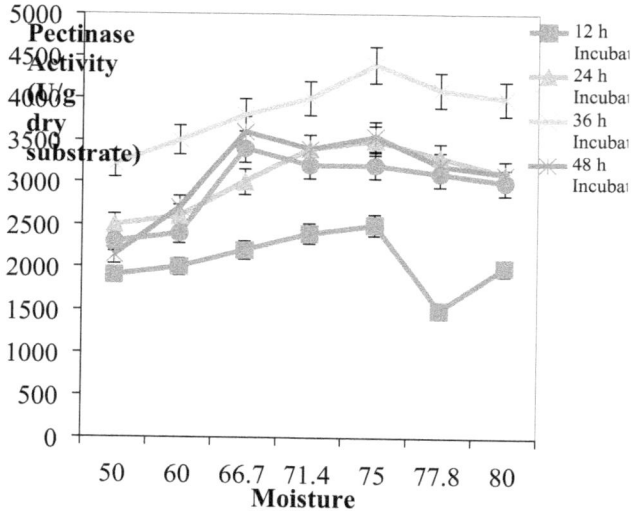

Figure 4.19: Effect of moisture content in solid-state medium (wheat bran) on alkaline pectinase production by *Bacillus tropicus* MCCC 1A01406 at 37°C, pH 9 under static conditions.

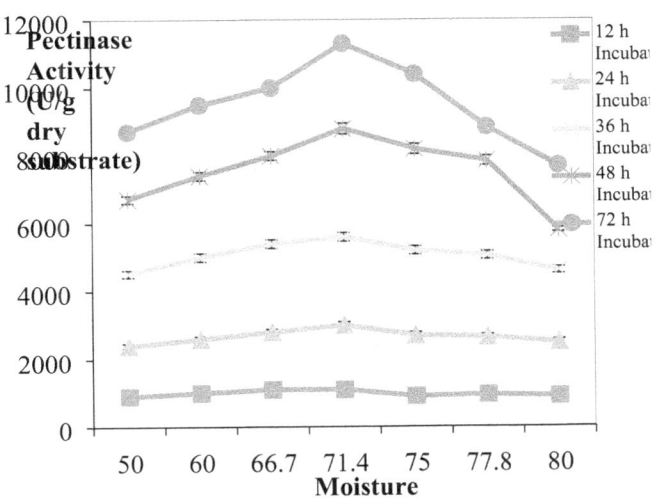

Figure 4.20: Effect of moisture content in solid-state medium (rice bran) on alkaline pectinase production by *Bacillus tropicus* MCCC 1A01406 at 37°C, pH 9 under static conditions.

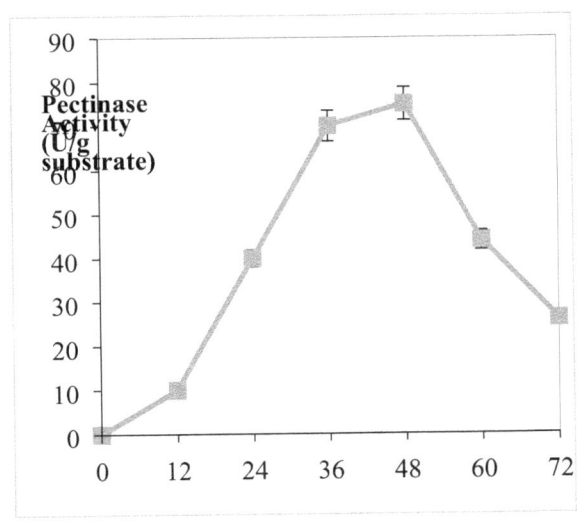

Figure 4.21: Solid state fermentation using apple pomace to produce alkaline pectinase from *Bacillus tropicus* MCCC 1A01406 at 37°C, pH 9 under static conditions.

4.5.2.2 Effect of salts

Various salts effect on pectinase synthesis by SSF was observed by supplementing distinct salts (final concentration, 1mM) into the wheat bran medium (moisture content, 71.4%). The addition of $CaCl_2 \cdot 2H_2O$ and $MgSO_4 \cdot 7H_2O$ increased pectinase synthesis by 28 and 11.6% respectively, although $CaCl_2 \cdot 2H_2O$ and $MgSO_4 \cdot 7H_2O$ do not affect pectinase production (Figure 4.22). H_3BO_3, $ZnCl_2$, KCl, and $NaCl$ decreased the enzyme synthesis to some extent.

4.5.2.3 Effect of different carbon sources

For the investigation of the effect of carbon supplementation on pectinase synthesis, various carbon sources (pectin, glucose, galactose, mannitol, sucrose, lactose, polygalacturonic acid, maltose, xylose, and sodium acetate, final concentration, 1% w/v), were supplemented into wheat bran medium (moisture content, 71.4%). Polygalacturonic acid, sodium acetate, pectin, and lactose were observed to increase pectinase production by up to 44% (Figure 4.23). And glucose, galactose, mannitol, sucrose, maltose, and xylose were found to inhibit pectinase production.

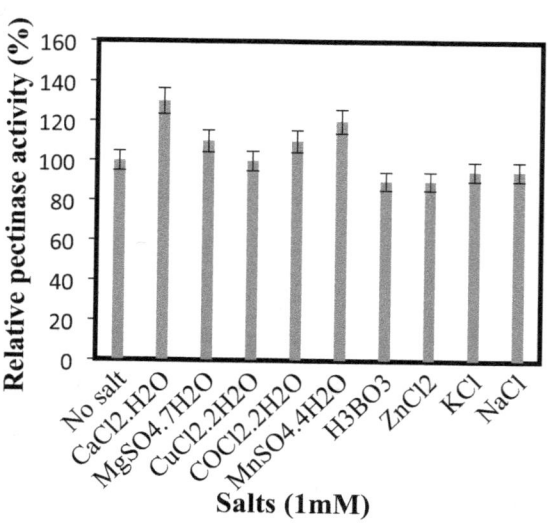

Figure 4.22: Production of alkaline pectinase by *Bacillus tropicus* **MCCC 1A01406 in a solid medium (wheat bran) using various salts (1Mm) at 37°C, for 36 h (moisture content 71.4%)**

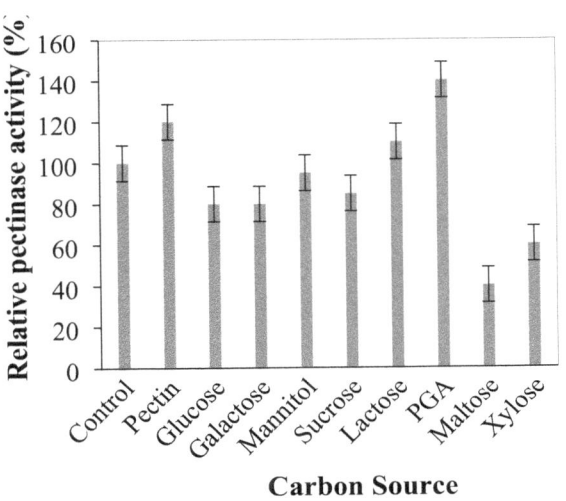

Figure 4.23: Alkaline pectinase production by *Bacillus tropicus* **MCCC 1A01406 in a solid medium (wheat bran) utilizing different carbon sources (1% w/v) at 37°C, for 36 h (moisture content 71.4%)**

4.5.2.4 Effect of various nitrogen sources

Various nitrogen source's effect on pectinase synthesis was observed by supplementation of these sources into wheat bran medium. Pectinase production was increased by Yeast extract, peptone, and ammonium chloride (1%, w/v) up to 24% (Figure 4.24). The addition of glycine, urea, and ammonium citrate inhibited pectinase production whereas; tryptone did not affect pectinase production in solid medium.

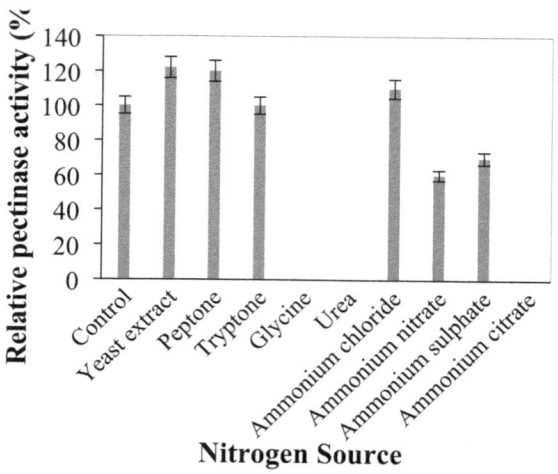

Figure 4.24: Alkaline pectinase production by *Bacillus tropicus* MCCC 1A01406 in a solid medium (wheat bran) utilizing different nitrogen sources at 37°C, for 36 h (moisture content 71.4%)

4.5.2.5 Effect of optimized components

Optimized components effect alone or in combinations on pectinase synthesis by *Bacillus* sp. P3 was observed. Various components supplied in combination of components like yeast extract (3%, w/w) + PGA; PGA + sodium acetate (3%, w/w), yeast extract + PGA + sodium acetate and PGA + sodium acetate + yeast extract + $CaCl_2.2H_2O$ enhanced pectinase synthesis from 43 to 70% based on the combination utilized (Table 4.8).

Table 4.8: Alkaline pectinase production by *Bacillus tropicus* MCCC 1A01406 in a solid medium (wheat bran) supplied with optimized components, at 37°C, for 36 h (moisture content 71.4%)

Components	Percent Relative activity
Control	100.00

Polygalacturonic acid	144.02
Sodium acetate	140.0
Yeast Extract	125.0
$CaCl_2.2H_2O$	166.0
PGA + YE	143.5
PGA + SA	148.4
PGA + $CaCl_2.2H_2O$	146.5
YE + SA	142.5
SA + $CaCl_2.2H_2O$	144.0
PGA + SA + $CaCl_2.2H_2O$	149.0
PGA + YE + $CaCl_2.2H_2O$	145.2
YE + SA + $CaCl_2.2H_2O$	162.0
PGA + SA + YE + $CaCl_2.2H_2O$	165.0

*PGA-Polygalacturonic acid; SA-Sodium acetate; YE-Yeast Extract

4.6 Purification of pectinase produced by *Bacillus tropicus* MCCC 1A01406

Purification of pectinase was achieved utilizing different steps. At first, the enzyme was partially purified by adding solid ammonium sulfate (0-40 and 40-100% cuts) to the cell-free supernatant. Pectinase activity was detected in 40-100% salt-saturated fraction. After dialysis of this fraction against 0.01 M Tri-HCl buffer (pH 9) under refrigeration conditions (4°C), this fraction was loaded onto anion exchanger, then DEAE- Sephacel column equilibrated with 0.01 M Tris-HCl buffer (pH 9). The proteins were eluted with NaCl gradient (0-1 M). Maximum proteins were displayed in two peaks; however, pectinase activity was observed only in the first peak fraction number 9.0 (Figure 4.25). Utilizing the DEAE-Sephacel chromatography, 68.2-fold purification of the enzyme was gained and its specific activity was observed to be 738.6 U/mg of protein. Fractions presenting pectinase activity were pooled, concentrated (utilizing an Amicon unit with 10 kDa molecular weight cut-off filter), and dialyzed with 0.01 M Tris-HCl buffer (pH 9). This protein sample was loaded on to Sephadex G-100 column and eluted with 0.01 M Tris-HCl buffer (pH 9). The pectinase activity was estimated in fractions 21-32 (Figure 4.26). These fractions were pooled and concentrated. This phase of purification yielded a 132.4-fold enhancement in pectinase activity and its specific activity was found to be 1466.2 U/mg proteins (Table 4.9).

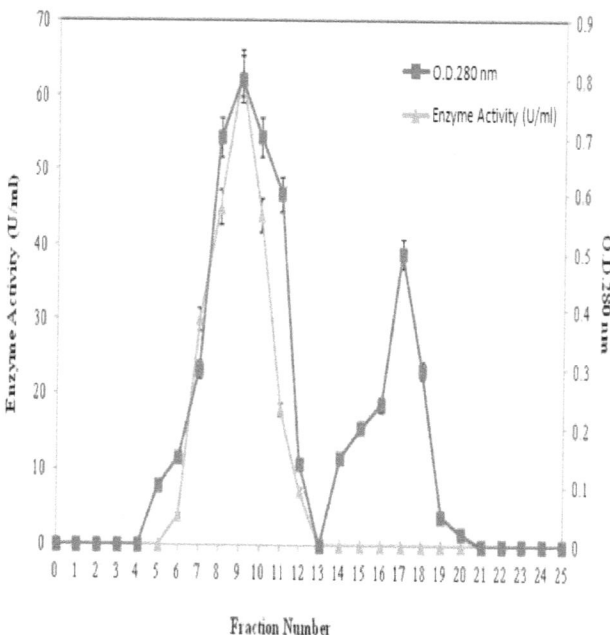

Figure 4.25: Purification of pectinase by ion-exchange chromatography using DEAE-Sephacel.

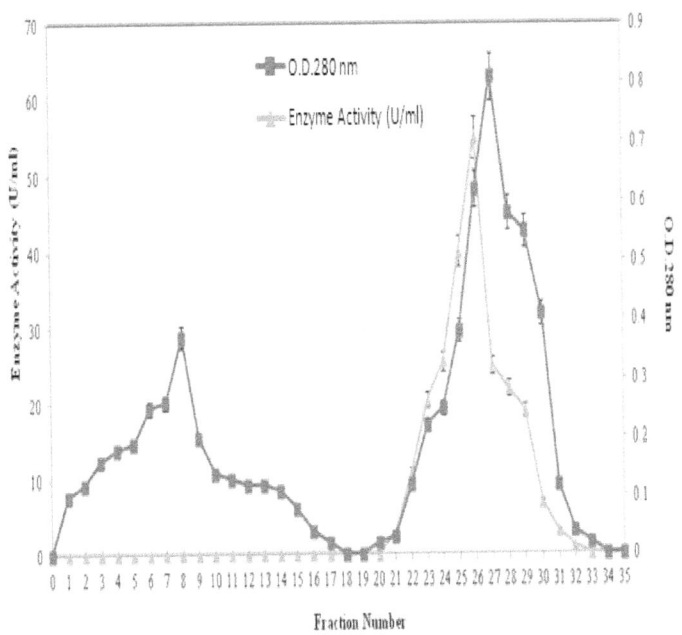

Figure 4.26: Elution profile of protein and pectinase activity on Sephadex G-100.

Table 4.9: Purification of pectinase by ammonium sulfate, DEAE Sephacel, and Sephadex G-100.

Purification steps	Volume of solution (ml)	Total enzyme activity (U)	Total protein (mg)	Specific activity (U/mg of protein)	Purification (fold)
Cell-free supernatant	250	4015.0	370.0	10.8	1.0
Ammonium sulphate precipitation (40-100% fraction)	50	1254.5	61.0	20.5	1.9
DEAE Sephacel	18.0	1122.7	1.52	738.6	68.2
Sephadex G100	10.0	513.2	0.35	1466.2	132.4

4.7 SDS-PAGE of purified pectinase

Purified pectinase was loaded and run on 10% SDS-PAGE. After Ag-staining the gel, a single band was found thus confirming complete purification of the enzyme. Using standard protein markers, the size of the purified enzyme was observed to be 98kDa from the graph between log molecular weight and migration distance.

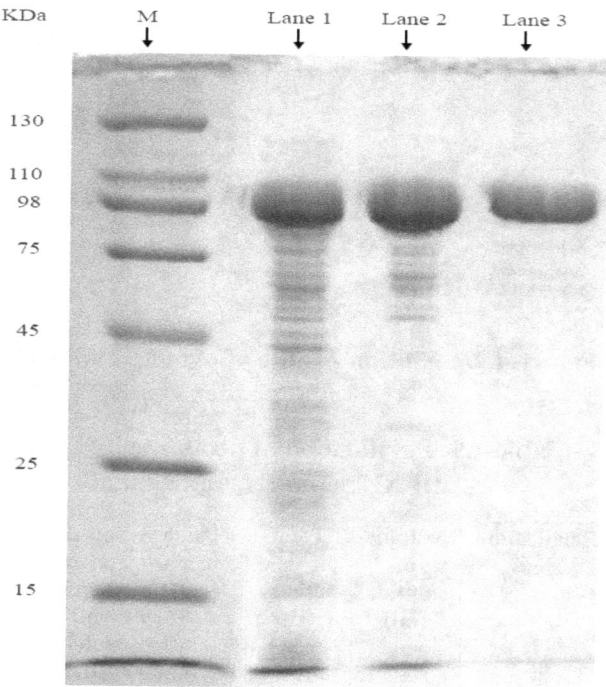

Figure 4.27 SDS-PAGE of pectinase on 10% gel at 50 V. Lane 1: molecular weight protein standards, lane 2: ammonium sulphate precipitated protein, lane 3: DEAE-Sephacel elute and lane 4: Sephadex G 100.

4.8 Characterization of purified pectinase produced by *Bacillus tropicus* MCCC 1A01406

4.8.1 Effect of pH on activity and stability of the enzyme

The purified enzyme was observed to be maximally active at pH 9.0 whereas almost a complete loss of activity was observed below pH 6.0 and above pH 12.0 (Figure 4.28). The pH stability of thepurified enzyme was checked at different pH values ranging from pH 4.0-12.0 and incubating the samples at room temperature for 4 h. Thereafter, the residual activity was determined at 60°C and pH value of 9.0. As shown in Figure 4.29, the enzyme was maximally stable at pH 9.0.

4.8.2 Effect of temperature on enzyme action and thermal stability

The effect of temperature at 45°C is expressed as 100% (Figure 4.30). To examine the thermal stability, this enzyme was kept at various temperatures (40-90°C) for 4 h. Then residual activity was estimated at pH 9.0 and temperature 45°C. As shown in Figure 4.31, the enzyme was maximally stable at 45°C. At 60°C the percent residual activity was reduced to 43%.

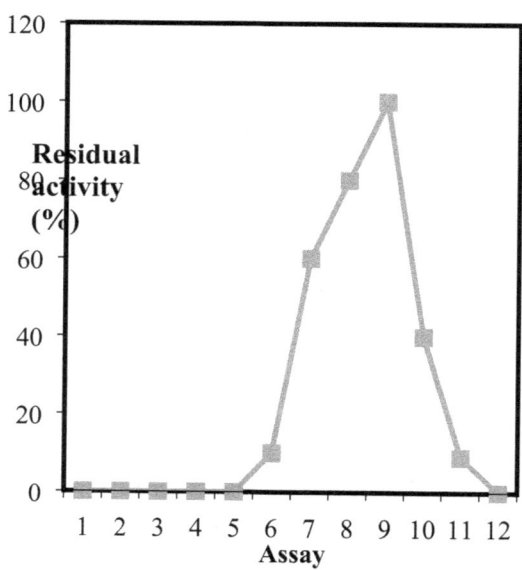

Figure 4.28: Assay pH effect on purified pectinase activity

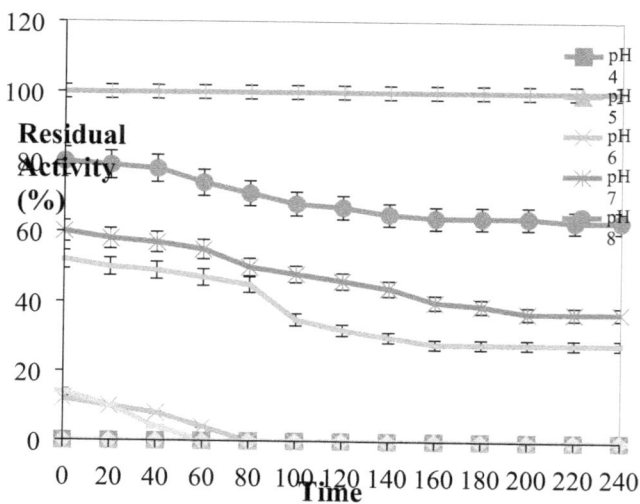

Figure 4.29: pH stability profile of purified pectinase from *Bacillus tropicus* MCCC 1A01406

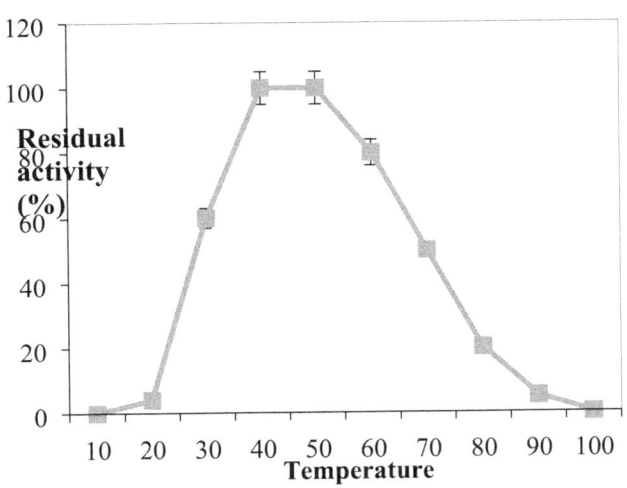

Figure 4.30: Assay temperature effect on purified alkaline pectinase activity

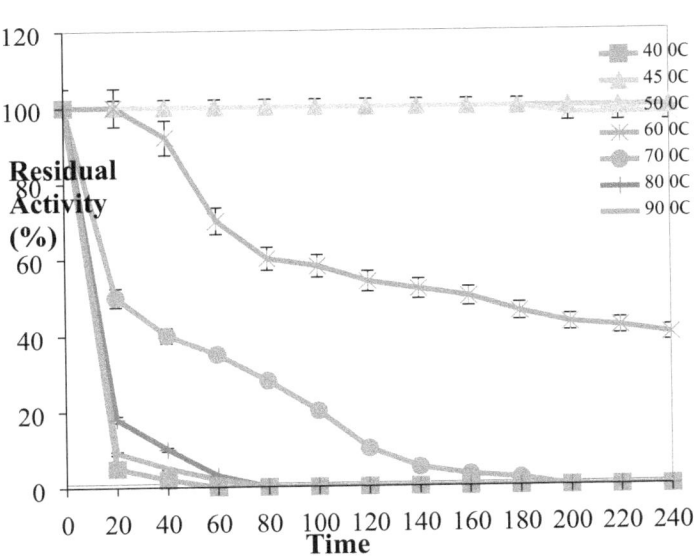

Figure 4.31: Thermostability profile of purified pectinase from *Bacillus tropicus* MCCC 1A01406.

4.8.3 Effect of chemical agents and surfactants on purified pectinase activity

Various chemical agents and surfactants (1mM, final concentration) effect on the purified pectinase activity was estimated by incubating the enzyme with these chemical agents at 45°C for 1 h(Table 4.10). However, urea, ascorbic acid, and cysteine inhibited pectinase activity. It was awesome to observe that the surface-active agents such as tweens(20, 40, 60, and 80), and SDS stimulated the pectinase activity up to 26%.

4.8.4 Effect of metal ions on pectinase activity

Table 4.11 shows the effect of different concentrations (0.5-10 mM) of metal ions on the purified enzyme activity. Ca^{2+} ions stimulated the pectinase activity by 44%, whereas Na^{2+}, Mn^{2+}, Pb^{2+}, Cu^{2+}, Ba^{2+} and Hg^{2+} decreased pectinase activity from 1-38% depending on the type of metal ion.

Table 4.10: Effect of different chemical agents and surfactants (1 mM) on purified pectinase activity.

Chemical agent/surfactant	Residual activity (%)
Control	100.0
Mercaptoethanol	140.0
Urea	100.0
Ascorbic acid	100.0
Glycine	84.6
Cysteine	100.5
EDTA	16.0
Tween 20	109.0
Tween 40	112.0
Tween 60	119.5
Tween 80	123.0
SDS (0.1%, w/v)	121.0
Triton - x - 100 (0.1 %, v/v)	126.0

Table 4.11: Metal ions and chelators effecton purified pectinase activity

Metal	Residualactivity (%) at various metal ion

ions	concentrations			
	0.5 mM	1.0 mM	5 mM	10 mM
Control	100.0	100.0	100.0	100.0
Ca^{2+}	128.0	144.8	144.0	144.0
Ba^{2+}	107.0	107.0	107.0	102.0
Hg^{2+}	73.0	74.0	69.0	67.0
Na^{2+}	102.0	104.0	98.0	96.5
Pb^{2+}	80.0	80.0	78.0	73.5
Cu^{2+}	98.7	98.2	98.0	97.6
Mn^{2+}	70.2	69.0	66.2	61.5

4.8.5 Kinetic parameters of pectinase activity

To ascertain whether the pectinolytic activity was due to hydrolysis or trans elimination, the end products of degradation of polygalacturonic acid by purified pectinase from *Bacillus* sp. P3 was analyzed by the thiobarbituric acid assay method [262]. The occurrence of the peak at 550 nm and its absence at 510 nm indicated the presence of only lyase activity and the absence of any hydrolase activity in the purified preparation (Figure 4.32). Therefore, the enzyme type present in the purified preparation was pectate lyase.

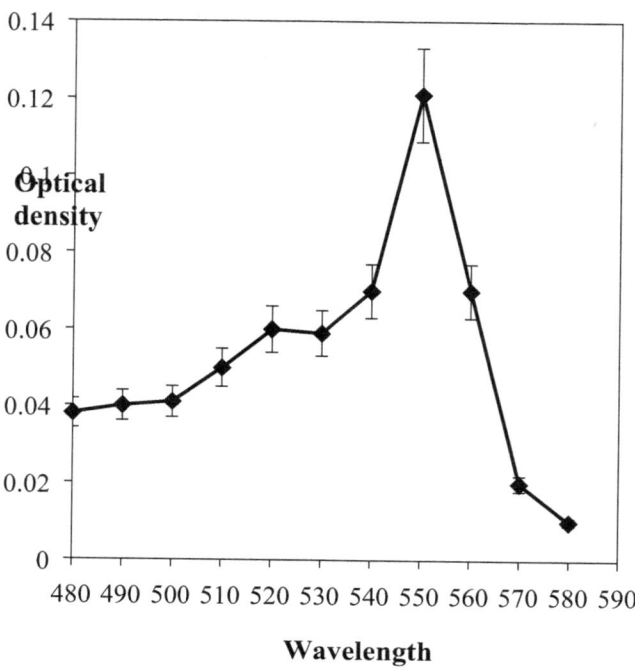

Figure 4.32: Absorption spectra of the degradation products of PGA at pH 9.0 acted on by alkaline pectinase containing pectic lyase activity.

4.8.6 Determination of Km and Vmax value

The pectinase from *Bacillus tropicus* MCCC 1A01406 exhibited a Km value of 4.8 mg/ml against PGA as substrate and the Vmax value of 390Umin^{-1} mg-1 proteins as determined using Lineweaver-Burk's plot (Figure 4.33).

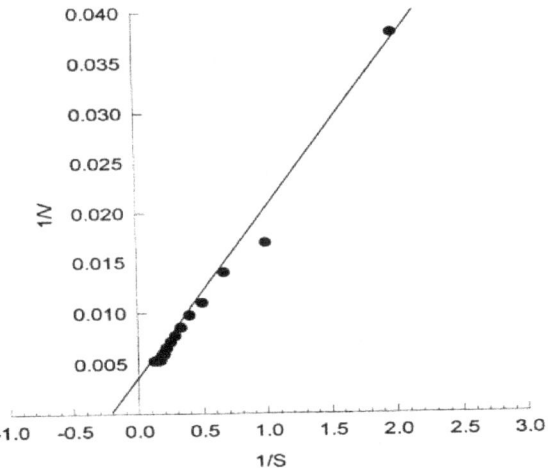

Figure 4.33: Lineweaver-Burk's plot showing Km and Vmax of purified pectinase.

4.9 Degumming of *Grewia optiva* bast fibers by RSM

4.9.1 Experimental Design

The experimental design techniques used for optimizing the process parameters for the maximum synthesis of galacturonic acid are studied with RSM and CCD. This system is optimizing the effective variables with a minimum number of experiments [268]. The four independent variables are studied at five levels (−2, −1, 0, +1, +2) (Table 4.12). A five-level and four-factor central composite rotatable design (CCRD) requiring a total of 30 runs for the determination of the experimental data. Reaction pH (X_1), temperature (X_2, °C), incubation time (X_3, hr), and pectinase concentration (X_4, g/l) are selected as independent variables and yield of galacturonic acid (mmole/g) as dependent variables (Table 4.13).

4.9.2 Statistical analysis

The effects of process parameters on response were analyzed by the RSM procedure to fit the second-order polynomial equation [268]. Statistica (StatSoft, Inc., USA) software was used to evaluate the experimental data. The basic form of the model equation is

$$Y = \beta_0 + \sum_{i=1}^{4} \beta_i X_i + \sum_{i=1}^{4} \beta_{ii} X_i^2 + \sum_{i=1}^{4} \sum_{j=i+1}^{4} \beta_{ij} X_i X_j \qquad (1)$$

Where Y represents the predicted response; β_0 is the mean effect, β_i, β_{ii}, β_{ij} are the regression coefficients for linear, quadratic, and cross-product coefficients and X_i and X_j represent the coded independent variables.

4.9.3 Analytical determination of degumming of fibers

The assessment of the degumming was done by considering the amount of galacturonic acid produced due to hydrolysis of pectic substances and finding the reduced weight of the fibers. For achieving these samples were withdrawn up to 24 hrs from the bacterial and chemical treatment process.

Table 4.12: Experimental design of four process variables in terms of coded values

Variables				
Coded levels	pH (X_1)	Temperature (°C) (X_2)	Incubation Time (Hrs) (X_3)	Enzyme Concentration (g/l) (X_4)

-2	7	27	1	0.5
-1	8	32	1.25	1
0	9	37	1.50	1.5
+1	10	42	2.15	2
+2	11	47	2.40	2.5

Table 4.13: Experimental design matrix and the experimental and predicted CCD design responses.

Run	pH (X_1)	Temperature (°C) (X_2)	Incubation Time (hr) (X_3)	Enzyme Concentration (g/l) (X_4)	Yield (mmole/g) Experimental	Yield (mmole/g) Predicted
1	8	32	1.25	2.0	460	461
2	8	32	2.15	1.0	311	310
3	8	42	1.25	1.0	352	353
4	8	42	2.15	2.0	458	459
5	10	32	1.25	1.0	300	296
6	10	32	2.15	2.0	412	411
7	10	42	1.25	2.0	416	414
8	10	42	2.15	1.0	303	302
9	9	37	1.5	1.5	450	448
10	9	37	1.5	1.5	445	448
11	8	32	1.25	1.0	350	354
12	8	32	2.15	2.0	423	421
13	8	42	1.25	2.0	485	487
14	8	42	2.15	1.0	319	320
15	10	32	1.25	2.0	402	399
16	10	32	2.15	1.0	302	303
17	10	42	1.25	1.0	285	283
18	10	42	2.15	2.0	440	437
19	9	37	1.5	1.5	449	448
20	9	37	1.5	1.5	447	448
21	7	37	1.5	1.5	430	427
22	11	37	1.5	1.5	320	325

23	9	27	1.5	1.5	412	412
24	9	47	1.5	1.5	433	432
25	9	37	0.6	1.5	391	391
26	9	37	2.4	1.5	413	412
27	9	37	1.5	0.5	184	183
28	9	37	1.5	2.5	422	423
29	9	37	1.5	1.5	448	448
30	9	37	1.5	1.5	447	448

4.9.4 Statistical analysis and model fitting

Response surface optimization is an added appropriate model in comparison to the traditional single-parameter optimization in that it preserves reaction time, space, and raw material. The levels of variables (pH, enzyme concentration, incubation time, and temperature) and the effects on the synthesis of galacturonic acid were determined by RSM. RSM using CCD consisting of 30 experiments was using different mixtures of variables. The results obtained from the CCD for the synthesis of galacturonic acid are shown in Table 4.12. From the statistical experimental design (Table 4.12), ANOVA, and Eq. (1), the second-degree polynomial response indicating the correlation among a synthesis of galacturonic acid and reaction variables are shown as in Eq. (2):

$$Y = 447.97 - 19.71 X_1 + 6.42 X_2 - 5.09 X_3 + 60.60 X_4 - 18.14 X_1^2 - 6.26 X_2^2 - 11.49 X_3^2 - 36.14 X_4^2 - 2.87 X_1 X_2 + 12.60 X_1 X_3 - 0.87 X_1 X_4 + 2.85 X_2 X_3 + 6.87 X_2 X_4 + 1.24 X_3 X_4 \quad (2)$$

Where Y is the galacturonic acid yield, and X_1, X_2, X_3, X_4 and X_5 are the reaction pH, incubation time and temperature, and enzyme concentration, respectively. A quadratic regression model (Eq. 2) describes the influence of the reaction variables on the synthesis of galacturonic acid was predicted. The reaction variables for galacturonic acid synthesis were imputed to the independent variables of reaction temperature, pH, incubation time, and enzyme concentration.

The statistical importance of the second-order polynomial equation was calculated by the F-test analysis of variance. The analysis of variance (ANOVA) is shown (Table 4.14) the significance of square term and first-order interaction terms of variables for the synthesis of galacturonic acid. The experimental and expected results match practically well with a high determination coefficient (R^2) was 0.98 for galacturonic acid yield. Thus, the model is acceptable for the estimation of such a reaction parameter [269].

The independent variables were more significant if larger F ratio and smaller P values [270]. The effects and interactions of variables were discussed from the Pareto chart shown in Fig. 4.34. The bar length in the Pareto chart is proportional to the absolute value of the regression coefficient. A variable can be significant if its related bar crosses this vertical line. From the Pareto chart (Fig. 4.35), linear coefficients term of reaction pH (X_1), temperature (X_2), incubation time (X_3) and enzyme concentration (X_4), quadratic coefficients term of reaction pH (X_1^2), temperature (X_2^2), incubation time (X_3^2) and enzyme concentration (X_4^2) and the first-order interaction terms like X_1X_2, X_1X_3, X_2X_3 X_2X_4, X_3X_4 were significant ($p < 0.05$) factors for the synthesis of galacturonic acid. The first-order interaction terms like X_1X_4 were the statistically insignificant terms ($p > 0.05$) for galacturonic acid synthesis.

Table 4.14: ANOVA for regression representing yield of galacturonic acid

Source of variance	Sum of squares	Degrees of freedom	Mean squares	F-ratio	P-value
X_1	9049.1	1	9049.08	3050.25	-
X_1^2	8976.6	1	8976.59	3025.82	-
X_2	960.4	1	960.39	323.73	0.000010
X_2^2	1071.4	1	1071.44	361.16	0.000007
X_3	629.4	1	629.44	212.17	0.000028
X_3^2	4088.4	1	4088.36	1378.10	-
X_4	85499.7	1	85499.65	28820.11	-
X_4^2	35623.5	1	35623.51	12007.93	-
$X_1 X_2$	132.3	1	132.25	44.58	0.001139
$X_1 X_3$	2711.5	1	2711.52	914.00	0.000001
$X_1 X_4$	12.2	1	12.25	4.13	0.097856
$X_2 X_3$	138.9	1	138.88	46.81	0.001018
$X_2 X_4$	756.2	1	756.25	254.92	0.000018
$X_3 X_4$	26.5	1	26.50	8.93	0.030487
Lack of fit	90.1	10	9.01	3.04	0.115955
Pure error	14.8	5	2.97		
Total	14536.4.7	29			

The effect and interaction of process parameters on the synthesis of galacturonic acid are shown in Figures 4.35 & 4.36.

Figure 4.35 & 4.36 (a) show the three-dimensional surface and contour plots of the effect of reaction pH and incubation time on galacturonic acid synthesis at a fixed temperature (37°C), enzyme concentration (1.5 g/l). From the plot, it can be observed that yield of galacturonic acid increased with an increase of reaction pH from 7 to 8, but yield decreased by further increasing of reaction pH. The incubation time effect on the synthesis of galacturonic acid is important. The yield increases with the incubation time increase from 1 to 1.25 hr, but the yield decreased by further increasing of incubation time. Wang et al. reported maximum bio scouring effect on cotton knitted fabrics with an alkaline pectinase at 1.25 hr reaction time [9].

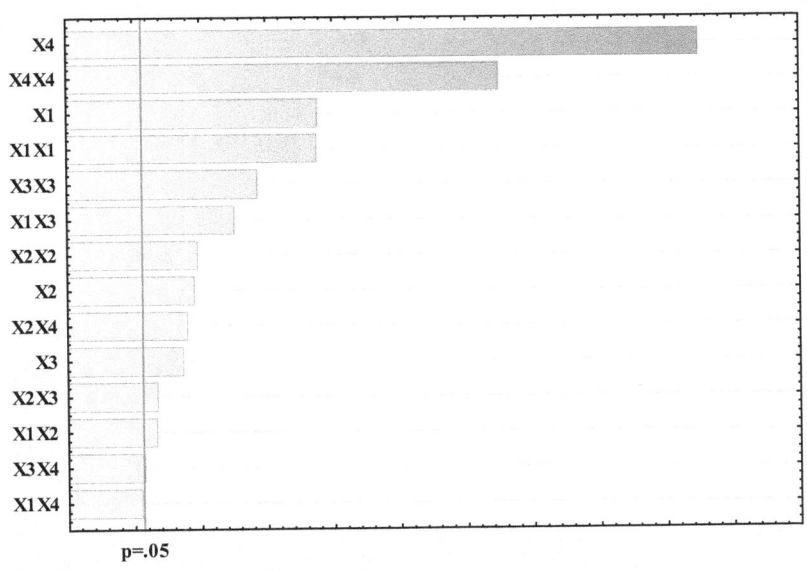

Figure 4.34: Pareto chart of standardized effect estimate of reaction conversion.

Figure 4.35 & 4.36 (b) exhibit the three-dimensional surface and contour plots at variousincubation times, also enzyme concentration at fixed reaction pH (9), temperature (37°C). The yield of galacturonic acid increased with the enzyme concentrationincrease from 0.5 to 2 g, drops from 2 to 2.5 g.

Figure 4.35 & 4.36(c) shows the three-dimensional surface and contour plots at varying incubation times, also reaction temperature at fixed reaction pH (9) and enzyme concentration (1.5g/l). The effect and interaction between incubation time and reaction temperature were significant on galacturonic acid synthesis. As can be observed, the galacturonic acid yield increases with the reaction temperature increase from 27 to 42°C and decreased promptly above 42°C.

(a) (b)

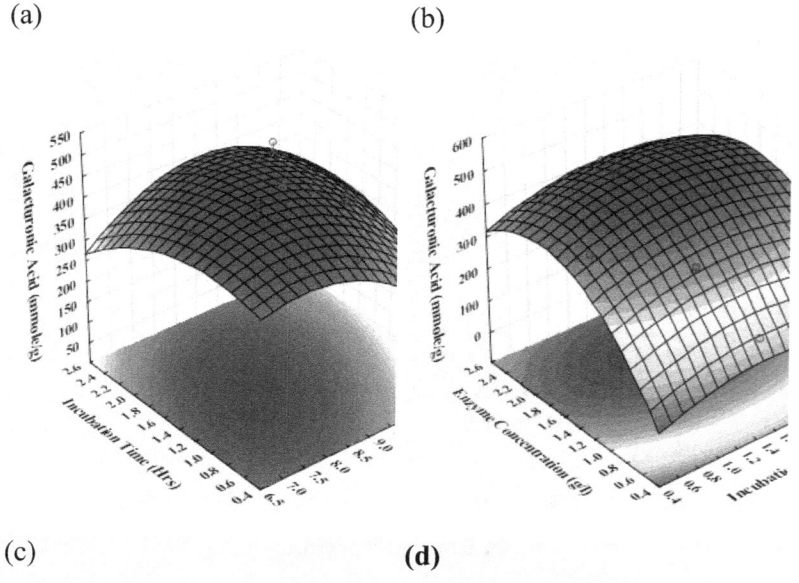

(c) (d)

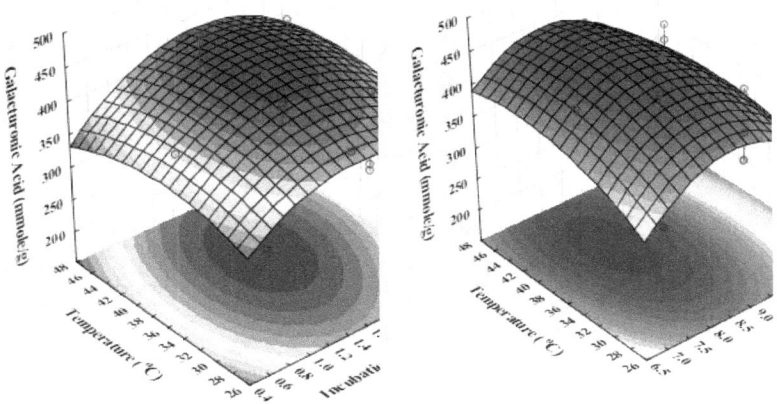

Figure 4.35: Response surface plot (3-D) showing the relation between the yield of galacturonic acid and: (a) At distinct reaction pH and incubation time (b) at distinctincubation time and enzyme (pectinase) concentration (c) at varying incubation time and temperature and (d) at different pH and reaction temperature.

| (a) | (b) |

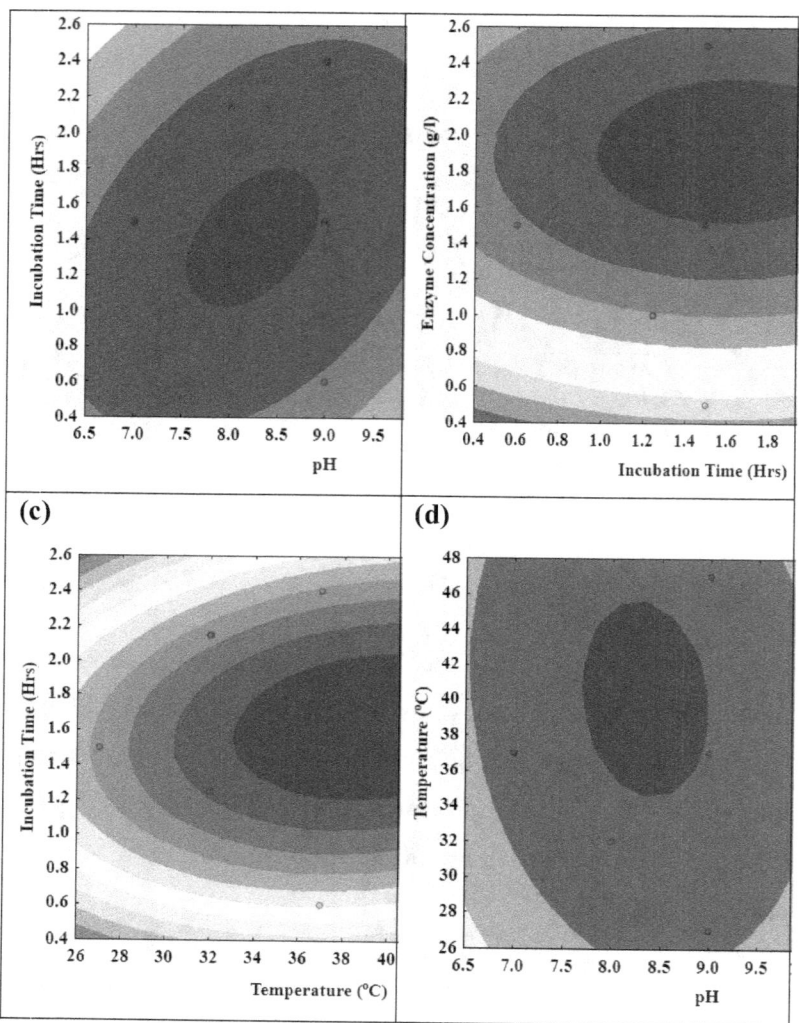

Figure 4.36: Contour plot exhibiting the effect of (a) the reaction pH and incubation time (b) incubation time and enzyme (pectinase) concentration (c) incubation time and temperature and (d) pH and reaction temperature on the galacturonic acid synthesis.

Figure 4.35 & 4.36 (d) shows the three-dimensional surfaces and contour plots of the effect of reaction pH and reaction temperature on the galacturonic acid synthesis. Enhancing the reaction pH (7 to 8) and reaction temperature (27 to 42°C) was increasing the yield of galacturonic acid, but more increase in reaction pH and the temperature was led to a decrease of galacturonic acid content.

4.9.5 Validation of Regression model

Four random statistical parameters are selected to calculate the galacturonic acid yield by regression Eq. 2. The CCD was used to find the exact values of the experimental conditions for the synthesis of galacturonic acid. From all experimental data, model predictions are indicating better compliance with the experimental data. The optimal experimental and predicted galacturonic acid of 487mmole/g was obtained under the optimum conditions of pectinase concentration of 2g/L, reaction pH of 8, and incubation time of 1.25hr at 42°C. These results show that the regression model was efficient to experimental and predicted conditions.

CHAPTER 5: DISCUSSION

Pectinases have found immense applications in the food sector as they effectively break down the pectin content of the plant material used for fruit juice synthesis, nectars, and purees. Generally, microbes produce two types of pectinases: acidic and alkaline. Over the last sixty years, the use of acidic pectinases has become a normal practice for clarification of fruit juices, as an aid to the extraction of juice and colored material, for the fruit, vegetable nectars, and purees synthesis [68]. Recently, alkaline pectinases are making encouraging inroads in the degumming and retting of fiber [93-94], pectic wastewater pretreatment from fruit juice industries [62, 224], also paper-making [229]. To date, not much innovation had been seen in these industrial sectors and relied on time-proven methods, which involved chemical treatment steps that result in environmental pollution. To overcome the problem of effluent disposal in an eco-friendly manner, biological processes are replacing chemical methods. The use of alkaline pectinases is one such example in this direction. The use of alkaline pectinases not only reduces environmental pollution it also improves the quality of the product. The cellulosic fibers obtained from fiber crops are the best textile material [93]. Conventionally these fibers are retrieved either by decortication or water retting of the stem. In both cases, bast fibers are freed from the surrounding tissues by maceration of the cortex. The water retting process takes a lot of time to complete and is weather-dependent. Moreover, the processed fibers still contain a large amount (20-40%) of gummy material made up of pectin, cellulose, and hemicellulose. On the other hand, this gummy material is removed in industries by treating the fibers with NaOH solution (12-20%) containing wetting and reducing agents. Sadly, the whole of the undesirable gummy material is not removed by this process. Therefore, a biotechnological degumming process including polysaccharide degrading microorganisms and their enzymes like pectinase has been found suitable [241].

We attempted to isolate a bacterium producing alkaline pectinase as logically fungi produce acidic pectinases, have a slower growth rate, produce lower levels of pectinase and the process for the separation of the enzyme from the biomass is complex as compared to bacteria. In the present study, a bacterial strain producing alkalophilic (pH 9.0), as well as thermotolerant (45°C) pectinase, was successfully isolated out of many bacterial strains tested. This strain was found to be producing a battery of other hydrolytic enzymes like lipase, amylase, cellulase, xylanase, and protease. This isolate's morphological, physiological, and biochemical characters were studied according to Bergey's Manual of Systematic Bacteriology [257]. Based on these characters this isolate was found to belong to Genus *Bacillus*. This bacterial isolate was named *Bacillus* sp. P3. After molecular characterization by using technique 16S rDNA, the isolate was identified as *Bacillus tropicus* MCCC 1A01406.Following the model given by Kelly and Fogarty [126], pectinase synthesis by **Bacillus tropicus** MCCC 1A01406 was observed to be essentially extracellular as this enzyme was found to be in the culture supernatant, and no pectinase activity was observed either in the cell lysate or in the cell washings of intact cells. This character of *Bacillus* sp. P3 is like the various reports in the literature [88, 213, and 246]. The pH optima of pectinase activity for the crude enzyme were found to be 9.0. The optimum temperature for the pectinase activity in the crude enzyme was 60°C, which is like the pectinase of *Bacillus* No. P-4-N, recorded by Horikoshi [66] has reported use of 1% (v/v) seed suspended in saline solution with the optical density of 10.0 at 600nm of *Bacillus* sp. RK9 cells for endo-PGL production. Whether all bacteria require calcium for pectinase production is difficult to say since several authors have used this during enzyme assay, without determining whether it was required. When calcium chloride was supplemented in a YEP medium to observe the effect of this ion on pectinase production from **Bacillus tropicus** MCCC 1A01406, it was observed that this ion at 1 mM concentration increased pectinase production more than three folds. This is clearly metal dependent pectinase, because EDTA is knocking down the activity.

To select an appropriate carbon source for maximum pectinase production various carbon sources (glucose, mannitol, glycerol, maltose, sucrose, lactose, cellobiose, fructose, xylose, sodium acetate, galactose, starch, arabinose, rhamnose, and galacturonic acid, 1%, w/v) were individually supplemented into YEP medium. It was observed that mannitol at 0.5% (w/v) level produced maximal activity (30.5 U/ml). All other carbon sources except galacturonic acid and sodium acetate were observed to inhibit pectinase production. The production of pectinase by different microbes shows induction and repression phenomena. Similar reports of catabolite repression exerted by carbon sources on the production of different pectic enzymes have been reported in the literature [119, 126, 246]. With most of these sugars used in the present study, the final pH of the medium remained below 8.0, whereas utilization of pectin, sodium acetate, galacturonic acid, and mannitol resulted in a final pH, ranging between 7-9. A complex nitrogen source (1%, w/v) was found to be required for pectinaseproduction as well as for biomass production by *Bacillus tropicus*MCCC 1A01406. High yields ofpectinase were obtained from **Bacillus tropicus** MCCC 1A01406with YE in the medium as a complexnitrogen source. Whereas glycine, urea, sodium nitrate, ammonium nitrateinhibited the growth [184]. Enzyme production by SSF usingbacterial spp. has been recorded for other enzymes like xylanase [203], amylase [205] but there is hardly any reportavailable on pectinase production by SSF utilizing any bacterial spp [163].

From the various substrates recorded in theliterature, wheat bran is considered best among all[132].In the present work, three substrates were utilized for the synthesis of pectinase, wheat bran was observed to be best for pectinase production by **Bacillus tropicus**MCCC 1A01406.These results are like those obtained with *Aspergillus niger*[21]. Thereafter, when wheat bran was suppliedwith various salts, carbon,and nitrogen sources, it was observed that $CaCl_2.2H_2O$ and $MgSO_4.7H_2O$(final concentration, 1mM) enhanced pectinase production up to 28% whereas, PGA,sodium acetate, and pectin increased the production up to 44%. In the present work, although polygalacturonic acid did not affect the production of pectinase in the SmF process, it enhanced pectinase production by44% in SSF. Similar was the case with sodium acetate. Glucose, galactose, sucrose,maltose, andxylose inhibited pectinase production in submerged as well as SSF process.Mannitol, whichenhanced pectinase production by 30% in the SmF process, did not affectthe pectinase productionmuch when supplemented into the solid medium. Whendifferent nitrogen sources were supplied into wheat bran medium, yeast extract andpeptone enhanced pectinase production up to 25%, whereas supplementation of 9 pl/mlmoistening agentenhanced pectinase production by almost 66%.

The pectinase from **Bacillus tropicus**MCCC 1A01406 was purified to homogeneity utilizing three steps: (i) ammonium sulfate precipitation, (ii) anion exchange chromatography, and (iii)gel filtration. When the cell-free supernatant exhibitingpectinase activity wasprecipitated with ammonium sulfate (0-40 and 40-100% salt saturation),pectinase activity was found in the 40-100% salt-saturated fraction. This fraction wasdialyzed with 0.01 M Tris-HCl buffer (pH 9) and loaded onto an anion exchanger(DEAE Sephacel). After elution with a 0-1M NaCl gradient, two peaks of proteins wereobserved, but only the second peak was associated with pectinase activity. This step of purification yielded 68.2 fold purification and the specific activity of the pectinase was 738.6 U/mg protein. In the third step, gel filtration (Sephadex G-100) of the pooled activefraction yielded a 132.4-fold increase in the pectinase purification and its specificactivity was observed 1466.2 U/mg protein. The fold increase in purification obtained in the presentstudy is much higher than that reported for **Bacillus purnilus** (16 fold) [122], and **Xanthomonas campestris** (66 fold) [123], but is lower than thatreported for **Clostridium multifermentans**(178 fold) (Millar and MacMillan, 1971) and **Bacillus** No. P-4-N (300 fold) [62]. Then this purified pectinase was electrophoresed on 10% SDS-PAGE a singleband was observed, and further, the molecular weight of the protein was determined to be 98 kDa by plotting a graph between migration distance and log molecular weight, whichis much higher than that of pectinase from **Fusariunoxysporium**f. sp. **Melonis**(58 kDa) [118]; **Bacillus pumilus** (20 kDa) [122]; **Bacillus stearothermophillus** (24 kDa) [85] and **Penicillium frequentans** (20 kDa) [91] reported in the literature.The purified pectinase was characterized concerning activity and stability atdifferent pH's and temperatures, the effect of chemical agents, surfactants, and metal ions,type of pectinolytic activity, and finally, determination of Km and **Vmax** values. The pHseems to affect the pectinase activity. Many fungal pectinases reported in the literatureare stable at low pH values, however pectinase from ***Bacillus tropicus*** MCCC 1A01406 was maximally active and stable under alkaline conditions of pH 8.0-9.0. A near-complete loss of pectinolytic activity at pH values less than 6.0 and more than 11.0 was observed. This range for pectinase stability is close to that reported for *Bacillus pumilus* (8.0-8.5) [122] and *Bacillus* strain SOI 13(8.4) [98].

The temperature activity and stability profile of the purified pectinase revealed that the enzyme was maximally active at moderately higher temperatures ranging from 40-60°C with the highest activity detected at 45°C. The pectinase was stable at 45°C for more than 4 h. This temperature optimum of ***Bacillus tropicus*** MCCC 1A01406 is under the temperature optima reported for *Erwinia carotovora* [9]; *Bacillus pumilus* [122], *Bacillus* GK-8 [88]; *Bacillus* sp.MG-cp-2 [100]; *Streptomyces* sp. QG-11-3 [16]. The effect of various chemical agents and surfactants on purified pectinase activity was studied. Mercaptoethanol stimulated pectinase activity up to 45%, whereas urea, ascorbic acid, cysteine, EDTA, and iodoacetic acid completely inhibited the enzyme activity. The inhibition of pectinase activity by iodoacetic acid indicated that cysteine residues are part of the catalytic site of pectinase. It was interesting to note that all the surface-active detergents i.e, tweens- (80, 60, 40, and 20), triton X-100, and SDS stimulated the pectinase activity from 7 to 26%. The reason is most probable that the surface-active agents might have increased the turnover number of alkaline pectinase by increasing the contact frequency between the active site of the enzyme and the substrate by lowering the surface tension of the aqueous medium. Like many pectinases reported in the literature, the enzyme in the present study was also activated (by 45%) by calcium ions. Significant inactivation of the pectinase activity was observed with Cu^{2+}, Mn^{2+}, Pb^{2+}, Hg^{2+}, and Ba^{2+}. The pectinase produced by ***Bacillus tropicus*** MCCC 1A01406 seemed to belong to the pectic lyasefamily based on the analysis of end products of PGA degradation by the thiobarbituric acidassay method [262]. The occurrence of a peak at 545-550 nm indicated pectin lyase activity, whereas the absence of a peak at 510 nm indicated the absence of hydrolaseactivity. Therefore, the pectinase activity in the purified enzyme was considered solelydue to trans-elimination action. Lyases that degrade pectic substances by trans eliminative mechanisms are widespread. They have been reported from several fungi [262], bacteria [119, 171].The lack of standardization in the substrate utilized and in the temperature and pH conditions during the assay makes it difficult to compare the Km values of pectinases from different microorganisms. With this consideration, Km values of 1.8 mg/ml for citrus pectin in enriched pectin lyase preparations of *A. niger* [246] have been observed. In bacteria, a Km value of 4.8 mg/ml using highly methoxylatedpectin has been described for *Erwinia chrysanthemi* [224, 225]. In the present study ***Bacillus tropicus*** MCCC 1A01406exhibited a Km value of 4.8 mg/ml against PGA and theVmax value of 390 U min' mg' protein. A Km value of 8.0 mg/ml against citrus pectin hasbeen reported for polygalacturonase from *Sclerotinia sclerotiorum* [110, 271]. However, Polizeli *et al.* (1991) [116] has reported a Km value of 5.0 mg/ml for polypectatehydrolysis by an extracellular polygalacturonase from *Neurospora crassa*. In the present study ***Bacillus tropicus*** MCCC 1A01406 as well as crude pectinase from this bacterium was used for the degumming of buelfiber crops *Bacillus tropicus* alkaline pectinase was used as a catalyst in the synthesis of galacturonic acid from beul (*Grewia Optiva*) bast fibers. RSM was used to optimize the reaction parameters for the synthesis of galacturonic acid and a second-order response model was calculated. Reaction pH, incubation time, temperature, and enzyme concentration were the factors affecting the synthesis of galacturonic acid. The highest yield of galacturonic acid was 485mmole/g. The optimum reaction conditions for galacturonic acid synthesis were obtained as incubation time of 1.25hr, reaction pH of 8, alkaline pectinase concentration of 2g/l, and temperature of 42^0C.

The percentage weight losses of blue fibers after treatment with pectinase from ***Bacillus tropicus*** MCCC 1A01406 were found to be 26.4%. The degumming in terms of pmoles of reducing sugar released revealed that both the fibers exhibited maximum degumming when chemically treated fibers were subsequently treated with pectinase from ***Bacillus tropicus*** MCCC 1A01406. The increase in GA content and a high reduction in fiber weight after combined treatment may be because chemical treatment results in unmasking of pectic sites present on the fibers and thereby resulting in easier access to pectinase for subsequent action. Although the reduction in the gum content of degumming of the buel fibers with enzymatic methods is being recorded for the first time by using RSM.

As the use of monocultures for the processing of these fibers has been suggested by different workers, ***Bacillus tropicus*** MCCC 1A01406, as well as pectinase produced by this bacterium, represent a good alternative. In the traditional process for extraction of these fibers little or no control is exerted over the development of microbial flora or pectinolytic enzyme production and the completion of the process. In the present study, these processes have been carefully judged and found to give fibers of better quality as compared to the conventional mixed consortia and chemical degumming methods [124,241]. Mild chemical treatment of the fibers followed by pectinase treatment, therefore, represents a good alternative to produce good quality fibers.

CHAPTER 6: SUMMARY AND CONCLUSIONS

1. Screening of various mesophilic bacterial isolates resulted in the isolation of a bacterium producing alkalophilic and thermotolerant pectinase.
2. This isolate was characterized as ***Bacillus tropicus*** MCCC 1A01406based on morphological, biochemical, and molecular analysis. It was found to be Gram-positive, rod-shaped (2.0-2.5 pm), spore-forming, aerobic, non-motile, catalase, and oxidase-positive bacterium.
3. ***Bacillus tropicus*** MCCC 1A01406also produced other hydrolytic enzymes such as lipase, amylase, xylanase cellulase, and protease but lacked chitinase and tannase.
4. Pectinase produced by ***Bacillus tropicus*** MCCC 1A01406was extracellular.
5. This strain grew well on most of the carbon sources used but produced acid only in glucose, fructose, sucrose, galactose, and maltose. The isolate grew well between pH

and temperature range of 5.0-11.0 and 25-45°C respectively. This isolate was capable of hydrolyzing tributyine, xylan, starch, casein, and carboxymethyl cellulose, but did not hydrolyze tannic acid, chitin, and urea.

6. Of the various complex media tested, yeast extract (1%, w/v) supplemented with pectin (0.25%, w/v), was found to support maximum pectinase production.

7. Production of this enzyme was inducible.

8. The optimum conditions for pectinase production in the presence of **Bacillus tropicus** MCCC 1A01406 were 37°C, pH 9.0, with shaking (250 rpm) for 16 h (with 2% v/v, inoculum size) in YEP medium.

9. To enhance pectinase production in liquid YEP medium, various salts, carbon, and nitrogen sources were supplemented in the medium. The suplimentation of $CaCl_2.2H_2O$ and $MgSO_4.7H_2O$ (1.0mM) enhanced pectinase production almost three folds. The addition of $CuCl_2.2H_2O$, $CoCl_2.2H_2O$, and $MnSO_4.4H_2O$ (final concentration, 1 mM) also enhanced pectinase production. Out of the various carbon, sources studied, mannitol (0.5%, w/v) enhanced pectinase production by almost 0.3 folds whereas, most of the other carbon sources used except polygalacturonic acid, inhibited the enzyme production. None of the different nitrogen sources (except yeast extract in the YEP) tested were found to enhance pectinase production.

10. Maximum pectinase production (55.7 U/ml) was obtained in yeast extract pectin medium supplemented with 1 mM $CaCl_2.2H_2O$ alone as none of the other optimized components alone or in combinations enhanced pectinase production as compared to $CaCl_2.2H_2O$.

11. The production of pectinase was scaled up by solid-state fermentation. In SSF, using wheat bran as the prime solid substrate, 4600 U/g dry substrates of pectinase were obtained at 75% moisture content.

12. When optimized components were added into the solid medium individually or in combinations, PGA + Neurobion was found to be the best combination for pectinase production as this combination enhanced pectinase production by 75% as compared to the control containing wheat bran medium alone.

13. Pectinase from **Bacillus tropicus** MCCC 1A01406 was purified to homogeneity by ammonium sulfate precipitation, ion-exchange chromatography, and gel filtration chromatography. Using DEAE Sephacel chromatography, 68.2-fold purification of the enzyme was achieved and its specific activity was found to be 738.6 U/mg protein. Further purification of this enzyme using Sephadex G-100, yielded 132.4-fold purification and the specific activity of the purified enzyme was 1466.2 U/mg protein.

14. SDS-PAGE on 10% gel revealed a single band of the size 98kDa.

15. The purified pectinase was optimally active at 60°C and stable at 45°C for more than 4 h. The pH optima of the purified enzyme were 9.0 at 60°C and the enzyme was stable at pH 9.0, room temperature for more than 4 h. Mercaptoethanol (1 mM) stimulated enzyme activity by 45% whereas, urea, ascorbic acid, glycine, cysteine, and EDTA inhibited the pectinase activity. The surface-active agents such as tweens (80, 60, 40, and 20), triton-x-100, and SDS stimulated pectinase activity from 7 to 26%

whereas, iodoacetic acid completely inhibited pectinase activity. The Ca ions stimulated enzyme activity by 44% however, Ag^{2+}, Cu^{2+}, Mn^{2+}, Zn^{2+}, Pb^{2+}, Ba^{2+}, and Hg^{2+} inhibited pectinase activity.

16. The pectinase produced by **Bacillus tropicus** MCCC 1A01406 seemed to belong to the pectic lyase family.
17. The purified enzyme exhibited Km and Vmax values of 4.8 mg/ml and 390 U ml' min respectively.
18. **Bacillus tropicus** MCCC 1A01406, as well as crude pectinase from this bacterium, was used for the degumming of buel fiber crops *Bacillus tropicus* alkaline pectinase was used as a catalyst in the synthesis of galacturonic acid from *Grewia Optiva* bast fibers. RSM was used to optimize the reaction parameters for the synthesis of galacturonic acid and a second-order response model was calculated. Reaction pH, incubation time, temperature, and enzyme concentration were the factors affecting the synthesis of galacturonic acid. The highest yield of galacturonic acid was 485mmole/g. The optimum reaction conditions for galacturonic acid synthesis were obtained as incubation time of 1.25h, reaction pH of 8, alkaline pectinase concentration of 2g/l, and temperature of 42 ^{0}C.

REFERENCES

[1] Ouattara HG, Reverchon S, Niamke SL, Nasser W (2011) Molecular identification and pectate lyase production by Bacillus strains involved in cocoa fermentation. *Food Microbiology* 28(1):1-8.

[2] Shieh M, Brown RL, Whitehead MP, Carey JW, Cotty PL, Cleveland TE, Dean RA (1997) Molecular genetic evidence

for the involvement of a specific polygalacturonase P2c, in the invasion and spread of Aspergillus flavus in cotton bolls. *Appl Environ Microbiol* 63:3548–3552.

[3] Patil R, Dayanand A (2006) Exploration of regional agrowastesfor the production of pectinase by *Aspergillus niger*. *Food Technol Biotechnol* 44(2):289–292.

[4] Stutzenberger F. (1992) Pectinase Production. Encyclopedia of Microbiology. *Lederberg J.* 3; 327-337.

[5] Botella, C., Diaz, A., De Ory, I., Webb, C. and Blandino, A., (2007) Xylanase and pectinase production by Aspergillus awamori on grape pomace in solid-state fermentation. *Process Biochem*, 42; 98–101.

[6] Kertesz ZI, (1951) The Pectic Substances, *New York Interscience*, 41, 648.

[7] May, CD, (1990) Industrial Pectins: Sources, Production and Applications, *Carbohydrate Polym*, 12, 79-99.

[8] Rolin C, Pectin, In: Industrial gums, RL Whistler and JN BeMiller (eds), 1993, 3rd edn. New York: *Academic Press*.

[9] Hadj-Taieb, N., Ayadi, M., Bouabdallah, F. and Gargouri, A., (2002) Hyperproduction of pectinase activities by a fully constitutivemutant (CT1) of Penicillium occitanis. *Enzyme Microb Technol*, 30(5): 662–666.

[10] Dhiman SS, Sharma j, Battan B. (2008) Pretreatment processing of fabric by alkalothermophilic xylanase from *Bacillus stearothermophilus* SDX. *Enzyme Microb Technol,* **43;** 262-269.

[11] Wang Q, Fan X-R, Hua Z-Z, Chen J (2007) Optimizing bioscouring condition of cotton knitted fabrics with an alkaline pectinase from Bacillus subtilis WSHB04-02 by using response surface methodology. *Biochemical Engineering Journal* 34: 107-113.

[12] McNeil, M.; Darvill, A. G.; Albersheim, P. (1980) *Plant Physiol.* 66; 1128–1134.

[13] Ranadive, A. and Haard, N. (1973). Chemical nature ofstone cells from pear juice. *J. Food Sci.*, 38: 331-333.

[14] Frollini E, Bartolucci N, Sisti L, Celli A (2013) Poly (butylene succinate) reinforced with different lignocellulosic fibers. *Ind Crops Prod* 45:160–169.

[15] Hoondal G, Tiwari R, Tewari R, Dahiya NBQK, Beg Q (2002) Microbial alkaline pectinases and their industrial applications: a review. *Applied microbiology and biotechnology* 59(4-5):409-418.

[16] Phutela U, Dhuna V, Sandhu S, Chadha BS (2005) Pectinase and polygalacturonase production by a thermophilic *Aspergillus fumigatus* isolated from decomposting orange peels. *Braz J Microbiol* 36:63–69.

[17] Jayani, R.S., Saxena, S. and Gupta, R., (2005). Microbial pectinolytic enzymes: a review. *Process Biochem*, 40: 2931-2944.

[18] Beg, Q. K., Bhushan, B., Kapoor, M. and Hoondal, G. S. (2000). Production and characterization of thermostable xylanase and pectinase from Streptomyces sp. QG-11-3. *J. Ind. Microbiol. Biotechnol.*, 24: 396-402.

[19] Luh, B. and Kean, C. (1975). Canning of fruit. In. Commercial Fruit Processing, ed. J. Woodroof and B. Luh. *AVI Publishing*, Westport, CT. pp. 11-23.

[20] Agarwal S, Yadav RD, Mahajan R. (2016) Synergistic effect of xylano-pectinolytic enzymes produced by a bacterial isolate in bleaching of plywood industrial waste. *J Clean Prod.*;118:229-33.

[21] Rehman HU, Qader SA, Aman A (2012) Polygalacturonase: production of pectin depolymerising enzyme from *Bacillus licheniformis* KIBGE IB-21. *CarbohydrPolym* 90(1):387-391.

[22] Kaur G, Sarkar BC, Sharma HK (2004) Production, characterization and application of a thermostable polygalcturonase of a thermophilic mouldSporotricum thermophile. *Bioresour Technol* 94:239–243.

[23] Ahlawat S, Battan B, Dhiman SS, Sharma J, Mandhan RP (2014) Production of thermostable pectinase and xylanase for their potential application in bleaching of kraft pulp. *J Ind Microbiol Biotechnol* **34(12)**:763–770.

[24] Walia A, Guleria S, Mehta P, Chauhan A, Prakash J. (2017) Microbial xylanase and their industrial application in pulp and paper biobleaching: a review. *3Biotech.*;7(11):2-12.

[25] Garg A, Singh A, Kaur A, Singh R, Kaur J, Mahajan R. (2016) Microbial pectinases: An eco-friendly tool of nature for industries. *3 Biotech.*6:1–13.

[26] Pilar B, Carman S, Tmaes G. (1999) Villa production of pectic enzymes in yeast. *Microbiol Lett.*;175:1-9.

[27] Azeri C, Tamer AU, Oskay M (2010) Thermoactive cellulase free xylanase production from alkaliphilic Bacillus strains using various agro-residues and their potential Inbiobleaching of kraft pulp. *Afr J Biotechnol* **9(1)**:63-72.

[28] Mellon JE, Cotty PJ (2004) Expression of pectinase activity among *Aspergillus flavus* isolates from southwestern and southeastern United States. *Mycopathologia* 157:337-338.

[29] Boccas F, Roussos S, Gutierrez M, Serrano L and Viniegra GG (1994) Production of pectinase from coffee pulp in solid-state fermentation system: selection of wild fungal isolate of high potency by a simple three-step screening technique. *J Food Sci Technol* **31(1)**; 22–26.

[30] Pawar, S.B., Shah, H.D., Andhorika, G.R., (2002). Man-Made Textiles in India, 45(4), 13.

[31] Sawada, K., Tokino, S., Ueda, M., & Wang, X. (2003). Bioscouring of cotton with pectinase enzyme. *Journal of the Society of Dyers and Colourists*, 114, 333–336.

[32] Godfrey, T. and West, S. (1996). Introduction to industrial enzymology. In. Industrial Enzymology, 2nd ed., eds. Godfrey, T. and West, S. *Stockholm Press*, New York. pp. 1- 17.

[33] Garg A, Singh A, Kaur A, Singh R, Kaur J, Mahajan R. (2016) Microbial pectinases: An eco-friendly tool of nature for industries. *3 Biotech.;*6:1–13.

[34] Amin F, Bhatti HN, Bilal M. (2019) Recent advances in the production strategies of microbial pectinases —A review. *Int. J. Biol. Macromol.*122:1017–26.

[35] Favela-Torress E, Aguilar C, Esquivel-Contreras CJ, Gustavo GV. (2003) Pectinase Enzyme technology. *Asiatech Publisher Inc.* Delhi.273–6.

[36] Lu H (2005). Insights into cotton enzymatic pretreatment. *Internat. Dyer* 190(3): 10-13.

[37] May CD. (1990) Industrial pectins: Sources, production, and applications. *Carbohydr. Polym.*; 12:79-99.

[38] Shet AR, Desai SV, Achappa S. (2018) Pectinolytic enzymes: classification, production, purification and

applications. *Res. J. Life Sci Bioinform Pharm Chem Sci.*4:337–48.

[39] Kavuthodi B, Sebastian D. (2018) Review on bacterial production of alkaline pectinase with special emphasis on *Bacillus* species, *Biosci. Biotech. Res. Comm.*11(1):18-30.

[40] Tapre AR, Jain RK. (2014) Pectinases: Enzymes for fruit processing industry. *Int. Food Res J.*21(2):447–53.

[41] Kaur A, Singh A, Patra A K, Mahajan R (2016) Cost effective scouring of flax fibers using cellulose-free xylo-pectinolytic synergism from bacterial isolate. *Journal of cleaner production,* 1-5.

[42] Rebello S, Anju M, Aneesh E M, Sindhu R, Binod P, Pandey A. (2017) Recent advancements in the production and application of microbial pectinases: An overview. *Rev Environ Sci Biotechnol.*16(3):381–94.

[43] Dinu D, Dan V (1994) Selection of *Aspergillus niger* strains producing pectic enzymes. *Analeleuniversitatii Biologie* 43:33-37.

[44] Federici, F. (1985). Production, purification and partial characterization of an endopolygalacturonase from *Crytococcusalbidus* var. alvidus. *Antonie van Leeuwenhoek*, 52: 139-150.

[45] Gainvors, A., Karam, N., Lequart, C. and Belarbi, A. (1994). Use of *Saccharomyces cerevisiae* for the clarification of fruit juices. *Biotech. Lett.,* 18: 1329-1334.

[46] Kawano, C. Y., Chellegatti, M. A. D. S. C., Said, S. and Fonseca, M. J. V. (1999). Comparative study of intracellular and extracellular pectinases produced by *Penicillium frequentans, Biotechnol. Appl. Biochem.,* 29: 133-140.

[47] Lu H (2005). Insights into cotton enzymatic pretreatment. *Internat. Dyer* 190(3): 10-13.

[48] Loera, O., Aguirre, J. and Viniegra-Gonzalez, G. (1999). Pectinase production by a diploid construct from two *Aspergillus niger* overproducing mutants. *Enz. Microb. Technol.*, 25: 103-108.

[49] Semenova MV, Grishutin SG, Gusakov AV, Okunev ON, Sinitsyn AP (2003) Isolation and properties of pectinases from the fungus *Aspergillus japonicas*. *Biochem* 68(5):559-569.

[50] Esquivel JCC, Hours RA, Voget CE, Mignone CF. *Aspergillus kawachii* produces anacidic pectin releasing enzyme activity. *J Biosci Bioeng* 1999; 88:48-52.

[51] Whitaker, J. R. (1984). Pectic substances, pectic enzymes and haze formation in fruit juices. *Enz. Microb. Technol.*, 6: 341-349.

[52] Iguchi, K. I., Kishida, M. and Sakai, T. (1996). Purification and characterization of three extracellular protopectinases with polygalacturonase activities from Trichosporonpenicillatum. *Biosci. Biotechnol. Biochem.*, 60: 603-607.

[53] Ali, S. B. R., Muthuvelayudham, R., &Viruthagiri, T. (2013). Statistical optimization of medium components for hemicellulase production using tapioca stem. *Journal of Microbiology, Biotechnology and Food Sciences*, 2(6), 2377–2382.

[54] Askar, A., Gierschener, K., Siliha, H. and El-Zoghbi, M. (1990). Polysaccharides and cloud stability of tropical nectars. XXth *International Federation of Fruit juice Producers Symposium*, Paris, pp. 207-223.

[55] Amorim, H. V. and Amorim, V. L. (1977). Coffee enzyme and coffee quality. In. Enzymes in Food and Beverage Processing, ed. Ori, R. and St. Angelo, *A. J. ACS Symposium Series,* 47, 27-56.

[56] Bailey, M. J. and Pessa, E. (1990). Strain and process for production of poylgalacturonase. *Enz. Microb. Technol.,* 12: 266-271.

[57] Anis, P., &Eren, H. (2002). Comparison of alkaline scouring of cotton vs. alkaline pectinase preparation. *AATCC Review,* 2; 22–26.

[58] Collmer, A., Ried, J, & Mount, M. (1988). Assay methods for pectic enzymes. *Methods in Enzymology,* 161, 329–335.

[59] Kashyap DR, Vohra PK, Chopra S, Tewari R (2001) Applications of pectinases in the commercial sector: a review. *Biores Technol* 77:215–227.

[60] Gummadi SN, Panda T (2003) Purification and biochemical properties of microbial pectinases: a review. *Process Biochem* **38:**987–996.

[61] Caffall, Kerry Hosmer, and Debra Mohnen. (2009).The Structure, Function, and Biosynthesis of Plant Cell Wall Pectic Polysaccharides. *Carbohydrate Research* 344 **(14);** 1879–1900.

[62] Horikoshi, Koki (1999). Alkaliphiles : Some Applications of Their Products for Biotechnology. 63 (4); 735–750.

[63] Willats WG, Knox JP, Mikkelsen JD (2006) Pectin: new insights into an old polymer are starting to gel. *Trends in Food Science & Technology* 17(3):97-104.

[64] Wilson, R. H.; Smith, A. C.; Kacurakova, M.; Saunders, P. K.; Wellner, N.; Waldron, K. W. (2000). *Plant Physiol.* 124; 397-405.

[65] Sakai T, Sakamoto T, Hallaert J, Vandamme J (1993) Pectin, pectinase and protopectinase: production, properties and applications. *Adv Appl Microbiol* 39:213–294.

[66] Horikoshi K (1990) Enzymes of alkalophiles. In: Fogarty WM, Kelly CT (eds) Microbial enzymes and biotechnology, 2nd edn. *Elsevier Applied Science*, London, pp 275–294.

[67] Mellon JE, Cotty PJ (2004) Expression of pectinase activity among *Aspergillus flavus* isolates from southwestern and southeastern United States. *Mycopathologia* 157:337-338.

[68] Zhou C., Ye J., XueY. and Ma Y., (2015) Directed evolution and structural analysis of alkaline pectate lyase from the alkaliphilic bacterium *Bacillus* sp. Strain N16-5 to improve its thermostability for efficient ramie degumming, *Appl Environ Microbiol,* **81**, 5714-5723.

[69] Meyer, L. H. (1987). Carbohydrates. In. Food Chemistry. *LBS Publishers and Distributors*, New Delhi, pp. 87-95.

[70] Fellows, P.J. and Voragen, J. T. (1986). Studies on the growth of Candida utilis on Dgalacturonic acid and the products of pectin hydrolysis. *Enz. Microb. Technol.*, 9: 537- 540.

[71] Chenoweth, W. L. and Levielle, G. A. (1975). In. Physiological Effects of Food Carbohydrates, eds. Jeans, A. and Hodge, *J. American Chemical Society*, Washington, D. C. pp. 312-324.

[72] Pilnik, W. and Voragen, A. G. J. (1970). Pectic substances and other uronides. In. Biochemistry of Fruits and their Products, Vol. I, ed. *Hulme, A. C. Academic Press,* London, pp. 53-88.

[73] Bechard, S. and McMullen, J. N. (1986). Pectin-gelation microglobules: Effect of a crosslinking agent (formaldehyde) or in vitro dissolution rate. *International J. Pharmacy,* 31: 91-98.

[74] Bhardwaj, T. R., Kanwar, M., Lai, R. and Gupta, A. (2000). Natural gums and modified natural gums as sustained-release carriers. *Drug Dev. Ind. Pharmacy.,* 26 (10): 1025- 1038.

[75] Polizeli, T. L. M. D., Jorge, J. A. and Terenzi, H. F. (1991). Pectinase production by Neurospora crassa: Purification and biochemical characterization of extracellular polygalacturonase activity. *J. Gen. Microbiol.,* 137: 1815-1823.

[76] Yadav S, Shastri NV (2007) Purification and properties of an extracellular pectin lyase produced by the strain of *Penicillium oxalicum* in solid-state fermentation. *Ind Jour of Biochem and Biophy* 44:247-251.

[77] Singh SA, Ramakrishna M, Rao AGA (1999) Optimization of downstream processing parameters for the recovery of pectinase from the fermented broth of *Aspergillus carbonarious. Process Biochem* 35:411–417.

[78] Yadav S, Yadav PK, Yadav D, Yadav KDS (2009) Pectin lyase: a review. *Process Biochemistry* 44(1):1-10.

[79] Li J, Zhou P, Liu H, Lin J, Gong Y, Xiao W, Liu Z (2014) Monosaccharides and ethanol production from

superfine ground sugarcane bagasse using enzyme cocktail. *Bioresources* 9(2):2529–2540.

[80] Sharma D.C, Satyanarayan T (2004) Production and application of pectinolytic enzymes of Sporotrichumthermopohile and Bacillus pumilus. In: Reddy MS, Khanna S (eds) Biotechnolgical approaches for sustainable development. *Allied Publishers*, India, pp 164–169.

[81] Zou M, Guo F, Li X, Zhao J, Qu Y (2014) Enhancing production of alkaline polygalacturonate lyase from *Bacillus subtilis* by fed-batch fermentation. *PloS one* 9(3):90392.

[82] Gupta VK, Gaur R, Gautam N, Kumar P, Yadav IJ and Darmwal NS (2009) Optimization of xylanase production from *Fusarium solani*F7. *Am J Food Technol* 4(1); 20-29.

[83] Demir H, Tarı C (2014) Valorization of wheat bran for the production ofpolygalacturonase in SSF of Aspergillus sojae. *Industrial Crops and Products* 54:302-309.

[84] Taskin M (2013) Co-production of tannase and pectinase by free and immobilized cells of the yeast *Rhodotorulaglutinis* MP-10 isolated from tannin-rich persimmon (Diospyros kaki L.) fruits. *Bioprocess and biosystems engineering* 36(2):165-172.

[85] Karbassi, A. and Vaughn, R. H. (1980). Purification and properties of poygalacturonic acid trans-eliminase from *Bacillus stearothermophilus*. *Can. J. Microbiol.*, 26: 377-384.

[86] Sakellaris, G., Nikolaropoulos, S. and Evangelopoulos, A. E. (1989). Purification and

characterization of an extracellular polygalacturonic acid from Lactobacillus plantarum strain BA II. *J. Appl. Bacteriol.*, 67: 77-85.

[87] Rijssel, M. W., Gerwig, J. G. J. and Hausen, T. A. (1993). Isolation and characterization of an extracellular glycosylated protein complex from Clostridium thermosaccharolyticum with pectin methylesterase and polygalacturonate hydrolase activity. *Appl. Environ. Microbiol.*, 59 (3): 828-836.

[88] Dosanjh, N. S. and Hoondal, G. S. (1996). Production of constitutive, thermostable, hyperactive exopectinase from Bacillus GK-8. *Biotechnol. Lett.*, 18: 1435-1438.

[89] Gainvors, A. and Belarbi, A. (1995). Detection methods for polygalacturonase producing strains of *Saccharomyces cerevisae*. *Yeast*, 10: 1311-1319.

[90] Aguillar, G. and Huitron, C. (1990). Constitutive exo-pectinase produced by Aspergillus sp. CH-Y-1043 on different carbon sources. *Biotechnol. Lett.*, 12: 655-660.

[91] Borin, M. D. F., Said, S. and Fonseca, M. J. V. (1996). Purification and biochemical characterization of an extracellular endopolygalacturonase from *Penicillium frequentans*. *J. Agric. Food Chem.*, 44: 1616-1620.

[92] Solis-Pereyra, S., Favela-Torres, E., Gutierrez-Rojas, M., Roussoss, S., SaucedoCastaneda, G., Gunasekran, P and Vinegara-Gonzalez, G. (1996). Production of pectinases by *Aspergillus niger* in solid state fermentation at high initial glucose concentrations. *World J. Microbiol. Biotechnol.*, 12: 257-260.

[93] Bruhlman F, Kim KS, Zimmerman W, Fiechter A (1994) Pectinolytic enzymes from actinomycetes for the

degumming of ramie bast fibers. *Appl Environ Microbio* 160:2107–2112.

[94] Cao J, Zheng L, Chen S (1992) Screening of pectinase producer from alkalophilic bacteria and study on its potential application in degumming of rammie. *Enzyme Microb Technol* **4**:1013–1016.

[95] Alana, A., Alkorta, I., Dominguez, J. B., Llama, M. J. and Serra, J. L. (1990). Pectin lyase activity in a Penicillium italicum strain. *Appl. Environ. Microbiol.*, 56 (12): 3755-3759.

[96] Magro, P., Varvaro, L., Chilosi, G., Avanzo, C. and Balestra, G. M. (1994). Pectinolytic enzymes produced by Pseudomonas syringaepv. *Glycinea. FEMS Microbiol. Lett.*, 117: 1-6.

[97] Wong, D. W. S. (1995). Pectic enzymes. In. Food Enzymes, Structure and Mechanism, eds. *Chapman and Hall publications*. Pp. 212-236.

[98] Nasser, W., Chalet, F. and Robert-Baudoudy, J. (1990). Purification and characterization of extracellular pectate lyase from *Bacillus subtilis. Biochimie*, 72: 689-695.

[99] Gainvors, A., Karam, N., Lequart, C. and Belarbi, A. (1994). Use of *Saccharomyces cerevisiae* for the clarification of fruit juices. *Biotech. Lett.*, 18: 1329-1334.

[100] Kapoor, M., Beg, Q. K., Bhushan, B., Dadhich, K. S. and Hoondal, G. S. (2000). Production and partial purification and characterization of a thermo-alkali stable polygalacturonase from *Bacillus* sp. MG-cp-2. *Process Biochem.*, 36: 467-473.

[101] Acuna-Arguelles, M. E., Gutierrerz-Rozas, M., Viniegra-Gonzalez, G. and Favela-Torres, E. (1995).

Production and properties of three pectinolytic enzymes produced by *Aspergillus niger. Appl. Microbiol. Biotechnol.*, 43: 808-814.

[102] Borin, M. D. F., Said, S. and Fonseca, M. J. V. (1996). Purification and biochemical characterization of an extracellular endopolygalacturonase from Penicillium frequentans. *J. Agric. Food Chem.*, 44: 1616-1620.

[103] Channe, P. S. and Shewale, J. G. (1995). Pectinase production by Sclerotium rolfsii: Effect of culture conditions. *Folia Microbiol.*, 40: 111-117.

[104] Marcus, L., Barash, I., Sneh, B., Koltin. Y. and Finker, A. (1986). Purification and characterization of pectolytic enzymes produced by virulent and hypovirulent isolates of Rhizoctonia solani KUHN. *Physiol. Mol. Plant Pathol.*, 29: 325-336.

[105] Al-Obaidi, Z. S., Aziz, G. M. and Al-Bakir, A. Y. (1987). Screening of fungal strains for polygalacturonase production. J. Agric. *Water Resour. Res.*, 6: 125-182.

[106] Sone, H., Sugiura, J., Itoh, Y., Izaki, K. and Takahashi, H. (1988). Production and properties of pectin lyase in Pseudomonas marginalis induced by mitomycin C. *Agri. Biol. Chem.*, 52: 3205-3207.

[107] Kester, H. C. M. and Visser, J. (1990). Purification and characterization of polygalacturonases produced by the hyphal fungus *Aspergillus niger. Biotechnol. Appl. Biochem.*, 12 : 150-160.

[108] Fonseca, M. J. V. and Said, S. (1995). Sequential production of pectinases by Penicillium frequentans. *World J. Microbiol. Biotechnol*, 11: 174-177.

[109] Zou M, Li X, Shi W, Guo F, Zhao J, and Qu Y (2013) Improved production of alkaline polygalacturonate

lyase by homologous overexpression pelA in Bacillus subtilis. *Process Biochemistry* 48(8):1143-1150.

[110] Riou, C., Freyssinet, G. and Fevre, M. (1992). Purification and characterization of extracellular pectinolytic enzymes produced by Sclerotinia sclerotorum. *Appl. Environ. Microbiol.*, 58: 578-583.

[111] Thomas, L., Larroche, C., Pandey, A., (2013). Current developments in solid-state fermentation. *Biochem. Eng. J.* 81, 146–161.

[112] Huang, L. K. and Mahoney, R. R. (1999). Purification and characterization of an endopolygalacturonase from Verticillumalbo-atrum. *J. Appl. Microbiol.*, 86: 145-156.

[113] Pardo, C., Lapena, M. A. and Gacto, M. (1991). Purification and characterization of an extracellular exopolygalacturonase from Geotrichum lactis. *Can. J. Microbiol.*, 37: 974- 977.

[114] Elegado, F. B. and Fujio, Y. (1994). Purification and some properties of endopolygalacturonase from Rhizopus sp. LKN. *World J. Microbiol. Biotechnol.*, 10: 256-259.

[115] Bailey, M. J. and Pessa, E. (1990). Strain and process for production of poylgalacturonase. *Enz. Microb. Technol.*, 12: 266-271.

[116] Polizeli, T. L. M. D., Jorge, J. A. and Terenzi, H. F. (1991). Pectinase production by Neurospora crassa: Purification and biochemical characterization of extracellular polygalacturonase activity. *J. Gen. Microbiol.*, 137: 1815-1823

[117] Federici, F. (1985). Production, purification and partial characterization of an endopolygalacturonase from

Crytococcusalbidus var. alvidus. *Antonie van Leeuwenhoek*, 52: 139-150.

[118] Martinez, M. J., Alconda, M. T., Guillrn, F., Vazquez, C. and Reyes, F. (1991) Pectic activity from *Fusarium oxysporiumf.* sp. melonis: Purification and characterization of an exopolygalacturonase. *FEMS Microbiol. Lett.,* 81: 145-150.

[119] Fogarty, M. W. and Kelly, C. T. (1983). Pectic enzymes. In. Microbial Enzymes and Biotechnology, ed. Fogarty, M. W. *Applied Science Publishers*, London and New York, pp. 131-182.

[120] Horikoshi, K. (1972). Production of alkaline enzymes by alkalophilic microorganisms. Part III. Alkaline pectinase of *Bacillus* No P-4-N. *Agr. Biol. Chem.,* 36 (2): 285-293.

[121] Nagel, W. C. and Vaughn, R. H. (1961). The characteristics of a polygalacturonase produced by *Bacillus polymyxa. Arch. Biochem. Biophys.,* 93: 344-352.

[122] Dave, B. A. and Vaughn, R.H. (1971). Purification and properties of a polygalacturonic acid trans-eliminase produced by *Bacillus pumilus. J. Bacteriol.,* 108: 166-174.

[123] Nasuno, S. and Starr, M. P. (1967). Polygalacturonic acid traneliminase of *Xanthomonas compestris. Biochem. J.,* 104: 178-184.

[124] Chesson, A. and Codner, R. C. (1978). Maceration of vegetable by a strain of *Bacillus subtilis. J. Appl. Bacteriol.,* 44: 347-364.

[125] Markovic, O. and Kohn, R. (1984). Mode of pectin deesterification by *Trichoderma reeseipectinesterase. Experimentia,* 40: 842-843.

[126] Kelly, C. T. and Fogarty, W. M. (1978). Production and properties of polygalacturonate lyase by an alkalophilic microorganism *Bacillus* sp. RK9. *Can. J. Microbiol.*, 29: 1164- 1172.

[127] Ward, O. P. and Fogarty, W. M. (1974). Polygalacturonide lyase production by *Bacillus sublilis* and *Flavobacterium pectinovorum. Appl. Microbiol.*, 27: 346-350.

[128] Nagel, W. C. and Wilson, T. M. (1970). Pectic acid lyase of Bacillus polymyxa. *Appl. Microbiol.*, 20: 374-383.

[129] Nasuno, S. and Starr, M. P. (1966). Pectic enzymes of *Pseudomonas marginalis. Phytopathology*, 56: 1414-1415.

[130] Couto SR, Sanroman MA (2006) Application of solid-state fermentation to food industry—A review. *Journal of Food Engineering* 76 (3):291-302.

[131] Kashyap DR, Vohra PK, Chopra S, Tewari R (2001) Applications of pectinases in the commercial sector: a review. *Bioresour Technol* 77(3): 215- 227.

[132] Pedrolli DB, Carmona EC (2010) Purification and characterization of the exopolygalacturonase produced by *Aspergillus giganteus* in submerged cultures. *J Ind Microbiol Biotechnol* 37(6):567-573.

[133] Ahmed A., Naseem M.K., Ahmed A. Khan S.A. and Sohail M. (2019) Optimization of pectinase production from *Geotrichumcandidum* AA15 using response surface methodology. *Pak. J. Bot.* 51 (2): 743-750.

[134] Viniegra-Gonzaleza GE, FT, Cristobal Noe Aguilarb, Sergio de Jesus Romero- Gomeza, Gerardo Diaz-Godinez, Christopher Augurd (2003) Advantages of fungal

enzyme production in solid state over liquid fermentation systems.*Biochemical Engineering Journal* 13:157–167.

[135] Kapoor M, Kuhad RC (2000) Improved polygalacturonase production from *Bacillus* sp. MG-cp-2 under submerged (SmF) and solid state (SSF) fermentation.Lett *Appl Microbiol* 34(5):317-322.

[136] Mohandas A, Sindhu Raveendran, Binod Parameswaran, Amith Abraham, Raj SR Athira, Anil Kuruvilla Mathew, Ashok Pandey (2018) Production of Pectinase from *Bacillus sonorensis* MPTD1. *Food Technol. Biotechnol* 56(1):110-116.

[137] Roy K, Dey S, Uddin M, Barua R, Hossain M (2018) Extracellular Pectinase from a Novel Bacterium *Chryseobacterium indologenes* Strain SD and Its Application in Fruit Juice Clarification. *Enzyme research*.

[138] Zhang J, Zhao L, Gao B, Wei W, Wang H, Xie J (2018) Protopectinase production by Paenibacilluspolymyxa Z6 and its application in pectin extraction from apple pomace. *Journal of Food Processing and Preservation* 42(1):13367.

[139] Irshad M, Anwar Z, Mahmood Z, Aqil T, Mehmmod S, Nawaz H (2014) Bioprocessing of agro-industrial waste orange peel for induced production of pectinase by Trichoderma viridi; its purification and characterization. *Turkish Journal of Biochemistry* 39(1):9-18.

[140] Zohdi NK, Mehrnoush A (2013) Optimization of extraction of novel pectinase enzyme discovered in red Pitaya (Hylocereuspolyrhizus) peel. *Molecule* 18(11):14366–14380.

[141] Handa S, Nivedita S, Shruti P (2016) Multiple Parameter Optimization for Maximization of Pectinase production by *Rhizopus* sp. C4 under Solid State Fermentation. *Fermentation* 2(10):2-9.

[142] Alimardani-Theuil P, Gainvors-Claisse A, Duchiron F (2011) Yeasts: An attractive source of pectinases—From gene expression to potential applications: A review. *Process Biochemistry* 46(8):1525-1537.

[143] Glinka EM, Liao YC (2011) Purification and partial characterization of pectin methylesterase produced by *Fusarium asiaticum. Fungal Biol* 115(11):1112- 1121.

[144] Padma PN, Anuradha K, Reddy G (2011) Pectinolytic yeast isolates for cold-active polygalacturonase production. *Innovative food science & emerging technologies* 12(2):178-181.

[145] Rehman HU, Qader SA, Aman A (2012) Polygalacturonase: production of pectin depolymerising enzyme from *Bacillus licheniformis* KIBGE IB-21. *Carbohydr Polym* 90(1):387-391.

[146] Saadoun I, Dawagreh A, Jaradat Z, Ababneh Q (2013) Influence of Culture Conditions on Pectinase Production by *Streptomyces* sp. (Strain J9). *International Journal of Life Science and Medical Research* 3(4):148-154.

[147] Ahmed I, Zia MA, Hussain MA, Akram Z, Naveed MT, Nowrouzi A (2016) Bioprocessing of citrus waste peel for induced pectinase production by *Aspergillus niger;* its purification and characterization. *Journal of Radiation Research and Applied Sciences* 9(2):148-154.

[148] Jahan N, Shahid F, Aman A, Mujahid TY, Qader SAU (2017) Utilization of agro waste pectin for the

production of industrially important polygalacturonase. *Heliyon* 3(6): 00330.

[149] Jacob N (2009) Biotechnology for agro-industrial residues utilization, Part IV: Enzymes Degrading Agro-Industrial Residues and Their Production, Chapter 21: Pectinolytic Enzymes.*Springer* India 383-396.

[150] Li Z, Bai Z, Zhang B, Xie H, Hu Q, Hao C, Zhang H (2005) Newly isolated *Bacillus gibsonii* S-2 capable of using sugar beet pulp for alkaline pectinase production. *World Journal of Microbiology and Biotechnology* 21(8-9):1483-1486.

[151] Ahlawat S, Dhiman SS, Battan B, Mandhan RP, Sharma J (2009) Pectinase production by *Bacillus subtilis* and its potential application in biopreparation of cotton and micropoly fabric. *Process Biochemistry* 44(5):521-526.

[152] Afifi MM (2011) Effective technological pectinase and cellulase by *Saccharomyces cervisiae* utilizing food wastes for citric acid production. *Life Science Journal* 8(2):405-13.

[153] Hsu, E. J. and Vaughn, R. H. (1969). Production and catabolite repression of the constitutive polygalacturonic acid trans-eliminase of Aeromonas liquefaciens. *J. Bacteriol.*, 4: 172-181.

[154] Yu P, Zhang Y, Gu D (2017) Production optimization of a heat-tolerant alkaline pectinase from *Bacillus subtilis* ZGL14 and its purification and characterization. *Bioengineered* 8(5):613-623.

[155] Moreno-Garcia J, Garcia-Martinez T, Mauricio JC, Moreno J (2018) Yeast Immobilization Systems for Alcoholic Wine Fermentations: Actual Trends and Future Perspectives. *Frontiers in Microbiology* 9:241.

[156] Angelim AL, Costa SP, Farias BCS, Aquino LF, Melo VM M (2013) An innovative bioremediation strategy using a bacterial consortium entrapped in chitosan beads. *Journal of environmental management* 127:10-17.

[157] Diaz-Godinez G, Soriano-Santos J, Augur C, Viniegra-Gonzalez G (2001). Exopectinases produced by Aspergillus niger in solid-state and submerged fermentation: a comparative study. *Journal of Industrial Microbiology and Biotechnology* 26(5):271-275.

[158] Gophanea S.R., Khobragadea C.N., and Jayebhayea S.G. (2016) Extracellular pectinase activity from *Bacillus Cereus* GC subgroup a: isolation, production, optimization and partial characterization. *Journal of Microbiology, Biotechnology and Food Sciences*, 6(2), 767-772.

[159] Berbegal C, SpanoG, Tristezza M, Grieco F, Capozzi V (2017) Microbial resources and innovation in the wine production sector. *South African Journal of Enology and Viticulture* 38(2):156-166.

[160] Pericin, D., Kevresan, S., Banka, L., Antov, M. and Skrinjar, M. (1992). Separation of the components of pectinolytic complex produced by Polyporussquamosus in submerged culture. *Biotech. Lett.*, 14 (2): 127-130.

[161] Goncalves DB, Teixeira JA, Bazzolli DMS, de Queiroz MV, de Araujo EF (2012) Use of response surface methodology to optimize production of pectinases by recombinant *Penicillium griseoroseum* T20. Biocatalysis and Agricultural *Biotechnology* 1(2):140-146.

[162] Almeida C, Branyik T, Moradas-Ferreira P, Teixeira J (2003) Continuous production of pectinase by immobilized yeast cells on spent grains. *Journal of bioscience and bioengineering* 96(6):513-518.

[163] Zhang J, Kang Z, Ling Z, Cao W, Liu L, Wang M, Chen J (2013) High-level extracellular production of alkaline polygalacturonate lyase in *Bacillus subtilis* with optimized regulatory elements. *Bioresource technology* 146:543-548.

[164] Membre, J. M. and Burlot, P. M. (1994). Effect of temperature, pH and NaCl on growth and pectinolytic activity of Pseudomonas marginalis. *Appl. Environ. Microbiol.*, 60 (6): 2017-2022.

[165] Abdulrachman D, ThongkredP, Kocharin K, Nakpathom M, Somboon B, Narumol N, Chantasingh D (2017) Heterologous expression of Aspergillus aculeatus endo-polygalacturonase in Pichia pastoris by high cell density fermentation and its application in textile scouring. *BMC biotechnology* 17(1):15.

[166] Lonesane, B. K. and Ghidyal, N. P. (1992). Solid Substrate Cultivation. In. Exoenzymes, eds. Doelle, H. W., Mitchell, D. A. and Rolz, C. E. *Elsvier Science Publishers Ltd.*, London. Pp. 191-209

[167] Barnby, F. M., Morpeth, F. F. and Pyle, D. L. (1990). Endopolygalacturonase production from Kluyveromycesmarxianus. I. Resolution, purification and partial characterization of the enzyme. *Enz. Microb. Technol.*, 12: 891-897.

[168] Ros, J. M.,:Nunez, J. H., Saura, D., Sameron, M. C. and Laencina, J. (1991). Production of endopolygalacturonase from Rhizopus nigricans to the evaluation of growth substrate. *Biotechnol. Lett.*, 13: 287-290.

[169] Sandri IG, Lorenzoni CMT, Fontana RC, da Silveira MM (2013) Use of pectinases produced by a new

strain of *Aspergillus niger* for the enzymatic treatment of apple and blueberry juice. *LWT-Food Science and Technology* 51(2):469-475.

[170] Kobayashi, T., Koike, K., Yoshimatsu, T., Higaki, N., Suzumatsu, A., Ozawa, T., Hatada, Y. and Ito, S. (1999). Purification and properties of a low molecular weight, high alkaline pectate lyase from an alkalophilic strain of Bacillus. *Biosc. Biotechnol. Biochem.*, 63: 56- 72.

[171] Chatterjee, A. K., Buchanan, G. E., Behrens, M. K. and Starr, M. P. (1979). Synthesis and excretion of polygalacturonic acid trans-eliminase in Erwinia, Yersinia, and Klebsiella species. *Can. J. Microbiol.*, 25 (1): 93-102.

[172] Berovick, M. and Ostroversnik, H. (1997). Production of Aspergillus niger pectinolytic enzymes by solid-state bioprocessing of apple pomace. *J. Bacteriol.*, 53: 47-53.

[173] Nedovic V, Gibson B, Mantzouridou TF, BugarskiB, Djordjevic V, Kalusevic A, Yilmaztekin M (2015) Aroma formation by immobilized yeast cells in fermentation processes. *Yeast* 32(1):173-216.

[174] Minussi, R. C., Coelho, J. L. C., Baracat-Pereira, M. C. and Silva, D. O. (1996). Pectin lyase production by Penicillium griseoroseum : Effect of tea extract, caffeine, yeast extract and pectin. *Biotechnol. Lett.*, 18 (11): 1283-1286.

[175] Membre, J. M. and Burlot, P. M. (1994). Effect of temperature, pH and NaCl on growth and pectinolytic activity of Pseudomonas marginalis. *Appl. Environ. Microbiol*, 60 (6): 2017-2022.

[176] Oriol, E., Raimbault, M., Roussos, S. and Viniegra, G. (1988). Water and water activity in the solid state

fermentation of cassava starch by Aspergillus niger. *Appl. Microbiol. Biotechnol*, 27: 498-503.

[177] Roussos, S., Olmos, A., Raimbault, M., Saucedo-Castaneda, G. and Lonsane, B. K. (1991). Strategies for large scale inoculum for SSF systems : Conidiospores of *Trichoderma harzianum. Biotechnol. Lett.*, 5: 415-420.

[178] Gessesse, A. and Mamo, G. (1999). High level xylanase production by an alkalophilic Bacillus sp. by using solid state fermentation. *Enz. Microb. Technol*, 25: 68-72.

[179] Pandey, A., Soccoi, C. R. and Mitchell, D. (2000). New developments in solid state fermentation: I. Bioprocesses and products. *Process Biochem.* 35: 1153-1169.

[180] Cavalitto, S. F., Areas, J. A. and Hours, R. A. (1996). Pectinase production profile of Aspergillusfoetidus in solid state cultures at different acidities. *Biotech. Lett.*, 18: 251 - 256.

[181] Fonseca, M. J. V., Spandaro, A. C. C. and Said, S. (1991). Separation of the components of pectolytic complex produced by Tubercularia vulgaris in solid state cultures. *Biotech. Lett.*, 13: 39-42.

[182] Antier, P., Minjares, A., Roussoss, S., Raimbault, M and Viniegra-Gonzales, G. (1993). Pectinase hyper-producing mutants of Aspergillus niger C28B25 for solid-state fermentation of coffee pulp. *Enz. Microb. Technol.*, 15: 254-260.

[183] Minjares-Carranco, A., Trejo-Aguilar, B. A., Aguilar, G. and Vniegra-Gonzalez, G. (1997). Physiochemical comparison between pectinase producing mutants of *Aspergillus niger* adapted either to solid-state

fermentation or submerged fermentation. *Enz. Microb. Technol.*, 21: 25-31.

[184] Patil R, Dayanand A (2006) Exploration of regional agrowastesfor the production of pectinase by *Aspergillus niger. Food Technol Biotechnol* 44(2):289–292.

[185] Janani L K, Kumar G, Rao KVB. (2011) Screening of pectinase producing microorganisms from agricultural waste dump soil. *Asian J Biochem Pharm Res* 2(1); 329-337.

[186] Jahan N, Shahid F, Aman A, Mujahid TY, Qader SAU (2017) Utilization of agro waste pectin for the production of industrially important polygalacturonase. *Heliyon* 3(6): 00330.

[187] Lane, A. G. (1979). Methane from anaerobic digestion of fruit and vegetable processing wastes. *Food Technol.* in Australia, 36(3), 125-7.

[188] Lequerica, J. L. &Lafuente, B. (1977), Aprovechamiento de subproductoscftricos II. Fermentaci6n enmedios semi-sdlidos de corteza de naranja por Candida utilis. *Rev. Agroq y Tec. de Aliment.*, 17(1), 71-7.

[189] Rodrfguez, J. A., Echevarria, J., Rodriguez, F. J., Sierra, N., Daniel, A. & Martinez, O. (1985). Solid state fermentation of dried citrus peel by Aspergillus niger. *Biotech. Lett.,* 7(8), 577-80.

[190] Mohandas A, Sindhu Raveendran, Binod Parameswaran, Amith Abraham, Raj SR Athira, Anil Kuruvilla Mathew, Ashok Pandey (2018) Production of Pectinase from *Bacillus sonorensis* MPTD1. *Food Technol. Biotechnol* 56(1):110-116.

[191] Padma PN, Anuradha K, Reddy G (2011) Pectinolytic yeast isolates for cold-active polygalacturonase

production. *Innovative food science & emerging technologies* 12(2):178-181.

[192] Palaniyappan, M., Vijayagopal, V., Viswanathan, R., Viruthagiri, T., (2009). Screening of natural substrates and optimization of operating variables on the production of pectinase by submerged fermentation using Aspergillus niger MTCC 281. *Afr. J. Biotechnol.* 8, 682–686

[193] Wu, P., Luo, F., Lu, Z., Zhan, Z., & Zhang, G. (2020). Improving the Catalytic Performance of Pectate Lyase Through Pectate Lyase/Cu3 (PO4) 2 Hybrid Nanoflowers as an Immobilized Enzyme. *Frontiers in Bioengineering and Biotechnology*, 8, 280.

[194] Nadar, S. S., & Rathod, V. K. (2019). A co-immobilization of pectinase and cellulase onto magnetic nanoparticles for antioxidant extraction from waste fruit peels. *Biocatalysis and agricultural biotechnology*, 17, 470-479.

[195] Mohammadi, M., Mokarram, R. R., Shahvalizadeh, R., Sarabandi, K., Lim, L. T., &Hamishehkar, H. (2020). Immobilization and stabilization of pectinase on an activated montmorillonite support and its application in pineapple juice clarification. *Food Bioscience*, 100625.

[196] Omojasola PF, Jilani OP (2008) Cellulose production by *Trichodermalongi, Aspergillus niger* and *Saccharomyces cerevisiae* cultured on waste materials from orange. *Paki J ofBio Sci* 11(20):2382-2388.

[197] Kohli, P., & Gupta, R. (2019). Application of calcium alginate immobilized and crude pectin lyase from Bacillus cereus in degumming of plant fibres. *Biocatalysis and Biotransformation*, 37(5), 341-348.

[198] Khan, M. M., Choi, Y. S., Kim, Y. K., &Yoo, J. C. (2018). Immobilization of an alkaline endopolygalacturonase purified from Bacillus paralicheniformis exhibits bioscouring of cotton fabrics. *Bioprocess and biosystems engineering*, 41(10), 1425-1436.

[199] Garg, G., Singh, A., Kaur, A., Singh, R., Kaur, J., & Mahajan, R. (2016). Microbial pectinases: an ecofriendly tool of nature for industries. *3 Biotech*, 6(1), 47.

[200] Chakraborty, S., Jagan Mohan Rao, T., & Goyal, A. (2017). Immobilization of recombinant pectate lyase from *Clostridium thermocellum* ATCC-27405 on magnetic nanoparticles for bioscouring of cotton fabric. *Biotechnology progress*, 33(1), 236- 244.

[201] Amin, F., Bhatti, H. N., Bilal, M., &Asgher, M. (2017b). Improvement of activity, thermostability and fruit juice clarification characteristics of fungal exopolygalacturonase. *International journal of biological macromolecules*, 95, 97.

[202] Amin, F., A. Mohsin, H. N. Bhatti and M. Bilal. 2020. Production, thermodynamic characterization, and fruit juice quality improvement characteristics of an Exopolygalacturonase from Penicillium janczewskii. *BBA - Proteins and Proteomics*, 1868,140379.

[203] Amin F, Bhatti HN, Bilal M, Asgher M (2017) Multiple parameter optimizations for enhanced biosynthesis of exopolygalacturonase enzyme and its application in fruit juice clarification. *Int J Food Eng* 13(2):256.

[204] Bilal M, Asgher M, Iqbal HM, Ramzan M (2017). Enhanced bio-ethanol production from old newspapers

waste through alkali and enzymatic delignification. *Waste and Biomass Valorization* 8(7):2271-2281.

[205] Djordjevic V, Willaert R, Gibson B, Nedovic V (2017) Immobilized yeast cells and secondary metabolites. *Fungal Metabolites* 599-638.

[206] Bilal M, AsgherM, Ramzan M (2015) Purification and biochemical characterization of extracellular manganese peroxidase from *Ganoderma lucidum* IBL-05 and its application. *Scientific research and Essays* 10(14):456-464.

[207] Bhardwaj V, Degrassi G and Bhardwaj R.K (2017) Microbial pectinases and their applications in industries: a review. *Int. Res. J. of Eng. And Tech (IRJET)* 4(08):829-836.

[208] Munir N, Asgher M, Tahir IM, Riaz M, Bilal M, Shah SA (2015) Utilization of agro-wastes for production of ligninolytic enzymes in liquid state fermentation by *Phanerochaete chrysosporium*-IBL-03. IJCBS 7:9-14.

[209] Deng, Z., Wang, F., Zhou, B., Li, J., Li, B., & Liang, H. (2019). Immobilization of pectinases into calcium alginate microspheres for fruit juice application. *Food hydrocolloids*, 89, 691-699.

[210] Favela-Torres E, Volke-Sepulveda T, Viniegra-Gonzalez G (2006) Production of Hydrolytic DepolymerisingPectinases. *Food Technology & Biotechnology* 44(2):221-227.

[211] Teixeira MF, Lima Filho JL, Duran N (2000) Carbon sources effect on pectinase production from *Aspergillus japonicus* 586. *Brazilian Journal of Microbiology* 31(4):286-290.

[212] Goncalves DB, Teixeira JA, Bazzolli DMS, de Queiroz MV, de Araujo EF (2012) Use of response surface methodology to optimize production of pectinases by recombinant *Penicillium griseoroseum* T20. *Biocatalysis and Agricultural Biotechnology 1*(2):140-146.

[213] Fang, G., Chen, H., Zhang, Y., & Chen, A. (2016). Immobilization of pectinase onto Fe3O4@ SiO2–NH2 and its activity and stability. *International journal of biological macromolecules*, 88, 189-195.

[214] Asgher M, Khan SW, Bilal M (2016) Optimization of lignocellulolytic enzyme production by *Pleurotuseryngii* WC 888 utilizing agro-industrial residues and bio-ethanol production. *Romanian Biotechnological Letters* 21(1):11133.

[215] Kester, H. C. M. and Visser, J. (1990). Purification and characterization of polygalacturonases produced by the hyphal fungus Aspergillus niger. *Biotechnol. Appl. Biochem.*, 12: 150-160.

[216] Kobayashi, T., Koike, K., Yoshimatsu, T., Higaki, N., Suzumatsu, A., Ozawa, T., Hatada, Y. and Ito, S. (1999). Purification and properties of a low molecular weight, high alkaline pectate lyase from an alkalophilic strain of *Bacillus. Biosc. Biotechnol. Biochem.*, 63: 56- 72.

[217] Miller, L., and MacMillan, J. D. (1971). Mode of action of pectic enzymes: II. Further purification of exopolygalacturonate lyase and pectinesterase from *Clostridium multifermentans. J. Bacterol.*, 102 (1): 72-78.

[218] Kertesz, Z. I. (1951). Pectin enzymes. In. The Enzymes, Vol. I., eds. Summer, J. B. and Myrback, K. *Academic Press Inc.*, New York. pp. 745-769.

[219] Junwei C, Weihua S, Yong P, Shuyun C (2000) High-producers of polygalacturonase selected from mutants resistant to rifampin in alkalophilic *Bacillus* sp. NTT33. *Enzyme Microb Technol* 27:545–548.

[220] Kutateladze, L., Zakariashvili, N., Jobava, M.,Urushadze, T., Khvedelidze,R., Khokhashvili, I., (2009) Selection of Microscopic Fungi - Pectinase Producers, *Bulletin of the Georgian National Academy of Sciences*, 3(1), 136-141.

[221] Sanchez S, Demain AL (2002) Review: metabolic regulation of fermentation processes. *Enzyme Microb Technol* 31:895–906.

[222] Kareem SO, Adebowale AA (2007) Clarification of orange juice by crude fungal pectinase from citrus peel. *Niger Food J* 25:130–137.

[223] Rajagopalan G, Krishnan C (2008) Immobilization of maltooligosaccharide forming a-amylase from Bacillus subtilis KCC103: properties and application in starch hydrolysis. *J Chem Technol Biotechnol* 83(11):1511–1517.

[224] Chesson A (1980) Maceration in relation to the post handling and processing of plant material. *J Appl Biotechnol* **48**:1–45.

[225] Tanabe H, Kobayashi Y, Akamatsu I (1986) Pretreatment of pectic wastewater from orange canning by soft-rot Erwinia carotovora. *J Ferment Technol* 64:265–268.

[226] Tanabe H, Kobayashi Y (1987) Plant tissue maceration caused by pectinolytic enzymes from Erwinia spp. under alkaline conditions. *Agric Biol Chem* 51(10):2845–2846.

[227] Soresen SO, Pauly M, Bush M, Skjot M, McCann MC, Borkhardt B, Ulvoskov P (2000) Pectin engineering: modification of potato pectin by in vivo expression of endo-1,4- b-D-galacturonase. Proc Natl Acad Sci USA 97:7639–7644 Soriano M, Diaz P, Pastor FIJ (2005) Pectinolytic systems of two aerobic sporogenous bacterial strains with high activity on pectin. *CurrMicrob* 50:114–118.

[228] Palomaki T, Saarilahti HT (1997) Isolation and characterization of new C-terminal substitution mutation affecting secretion of polygalacturonases in *Erwinia carotovora* ssp. carotovora. *FEBS Lett* 400:122–126.

[229] Holbom B, Ekman R, Sjoholm R, Eckerman C, Thornton J (1991) Chemical changes in peroxide bleaching of mechanical pulps. *Das Papier A* 45(10):V16–V22.

[230] Reid, I., & Ricard, M. (2000). *Enzyme and Microbial Technology*, 26, 115 –123.

[231] West S (1996) Olive and other edible oils. In: Godfrey T, West S (eds) Industrial enzymology, 2nd edn. *Stockholm Press*, New York, pp 293–300.

[232] Perrone G, Mule` G, Susca A, Battilani P, Pietri A, Logrieco A (2006) Ochratoxin A production and AFLP analysis of Aspergillus carbonarius, Aspergillus tubingensis, and Aspergillus niger strains isolated from grapes in Italy. *Appl Environ Microbiol* 72:680–685.

[233] Maiorano AE, Schmidell W, Ogaki Y (1995) Short communication: determination of the enzymatic activity of pectinases from different microorganisms. *World J Microbiol Biotechnol* 11:355–356.

[234] Gunmadi, S.N. and Punda, T., 2003, Purification and biochemical of microbial pectinases—a review. *Process Biochem*, **38**; 987–996.

[235] Alkorta I, Garbisu G, Llama MJ, Serra JL (1998) Industrial applications of pectic enzymes: a review. *Process Biochem* 1:21–28.

[236] Kohli, P., & Gupta, R. (2019). Application of calcium alginate immobilized and crude pectin lyase from Bacillus cereus in degumming of plant fibres. *Biocatalysis and Biotransformation*, 37(5), 341-348.

[237] Liu, M., Fernando, D., Daniel, G., Madsen, B., Meyer, A. S., Ale, M. T., &Thygesen, A. (2015). Effect of harvest time and field retting duration on the chemical composition, morphology and mechanical properties of hemp fibers. *Industrial Crops and Products*, 69, 29-39.

[238] Kirk TK, Jefferies TW (1996) Role of microbial enzymes in pulp and paper processing. In: Jefferies TW, Viikari L (eds) Enzymes for pulp and paper processing. ACS symposium series. *American Chemical Society*, Washington, pp 1–14.

[239] Reid I, Ricard M (2000) Pectinase in paper making: solving retention problems in mechanical pulp, bleached with hydrogen peroxide. *Enzyme Microb Technol* 26:115–123.

[240] Bajpai, P. (1999). *Biotechnology Progress*, 15, 147–157.

[241] Ahlawat S, Battan B, Dhiman SS, Sharma J, Mandhan RP (2007) Production of thermostable pectinase and xylanase for their potential application in bleaching of kraft pulp. *J Ind Microbiol Biotechnol* 34:763-770.

[242] Baracat-Pereira, M. C., Vanetti, M. C. D., Aruojo, E. F. D., Silva, D. O. (1993). Partial characterization of Aspergillus fumigatus polygalacturonase for the degumming of natural fibers. *J. Ind. Microbiol.*, 11: 139-142.

[243] Bhattacharya, S. K. and Paul, N. B. (1976). Susceptibility of ramie with different gum contents to microbial damage. *Curr. Sci.*, 45: 417-418.

[244] Gurucharanam, K. and Deshpande, K. S. (1986). Polysaccharases of Curbularialunata: Use in degumming oframie fibers. *Ind. J. Phytopathol.*, 39 (3): 385-389.

[245] Sharma, H. S. S. (1987). Enzymatic degradation of residual non-cellulosic polysaccharides present on dew-retted flax fibers. *Appl. Microbiol. Biotechnol*, 26: 358-362.

[246] Henriksson, G., Akin, D. E., Hanlin, R. T., Rodriguez, C., Archibald, D., Rigsby, L. L. and Eriksson, K. E. L. (1997). Identification and retting efficiencies of fungi isolated from dew-retted flax in the United States and Europe. *Appl. Environ. Microbiol*, 63: 3950- 3956.

[247] Rodrigues PR, Silverio TAB, Druzian JI. Microbial Synthesis and Characterization of Biodegradable Polyester Copolymers from *Burkholderia Cepacia*and*Cupriavidus Necator* Strains Using Crude Glycerol as Substrate. *Braz Arch Biol Technol*. 2019;62:e19170498.

[248] Battan B, Dhiman SS, Ahlawat S, Mahajan R, Sharma J. Application of thermostable xylanase of *Bacillus pumilus*in textile processing. *Indian J.Microbiol.* 2012;52:222-9.

[249] Loow YL, Wu TY, Tan KA, Lim YS, Siow LF, Jahim JM, Mohammad AW, Teoh WH. Recent advances in the application of inorganic salt pretreatment for

transforming lignocellulosic biomass into reducing sugars. *J Agric Food Chem.* 2015;63(38):8349-63.

[250] Dutta, AK, Ghosh BL, Aditya RN. The enzymatic softening and upgrading of lignocellulosic fibers: part III: pre-treatment of jute with enzymes for fine-yarn spinning. *J. Text. Inst.* 2000;91:28-34.

[251] Zolriasatein AA, Yazdanshenas ME. Changes in composition, appearance, physical, and dyeing properties of jute yarn after bio-pretreatment with laccase, xylanase, cellulase, and pectinase enzymes. *J Text Inst.* 2014;105:609-19.

[252] Zetelaki, K. (1976). Optimal carbon source concentration for the pectinolytic enzyme formation of *Aspergillus. Proc. Biochem.*, 11: 11-18.

[253] Rombouts, F. M. and Thibault, J. F. (1986). Enzyme and chemical degradation and the fine structure of pectins from sugar-beet pulp. *Carboh. Res.*, 154: 189-203.

[254] Godfrey, T. and West, S. (1996). Introduction to industrial enzymology. In. Industrial Enzymology, 2nd ed., eds. Godfrey, T. and West, S. *Stockholm Press*, New York. pp. 1- 17.

[255] Soares MMCN, Da Silva R, Carmona EC, Gomes E. (2001) Pectinolytic enzyme production by *Bacillus* species and their potential application of juice extraction. *World J Microbiol Biotechnol.*17:79-82.

[256] Miller GL. Anal. Chem. Use of Dinitrosalicylic acid reagent for determination of reducing sugar. 1959; 31:426-428.

[257] Lowry OH, Roserbrough NJ, Farr AL, Randall RJ. (1951) Protein measurement with the Folin phenol reagent. *J Biol Chem.*193:265-75.

[258] Sneath, P. H. A. (1994). Endospore forming Gram positive rods and cocci. In. Bergey's Manual of Systematic Bacteriology, 9th ed., eds. Hensyl, W. M. Williams and *Wilkins Publishers*, USA. pp. 1104-1139.

[259] Chen W, Kuo T. A (1993) simple and rapid method for the preparation of gram-negative bacterial genomic DNA. *Nucleic Acids Res*.21:2260.

[260] Ansari A, Aman A, Siddiqui NN, Iqbal S, Qader SAU. (2012) *Pak J Pharm Sci*.25:195-201.

[261] Tamura K, Dudley J, Nei M, Kumar S. (2007) MEGA4: Molecular evolutionary genetics analysis (MEGA) software version 4.0. *Mol Biol Evol*.24:1596-9.

[262] Laemmli, U. K. (1970) Cleavage of structural proteins during the assembly of head bacteriophage T4. *Nature*, 22: 680-685.

[263] Sherwood, R. T. (1966). Pectin lyase and polygalacturonase production by Rhizoctonia solani and other fungi. *Phytopathol.*, 36: 279-286.

[264] Luedeking R, Piret EL. A kinetic study of the lactic acid fermentation: batch process at controlled Ph. *J Biochem Microbiol Technol Eng*. 1959; 1:393-412.

[265] Holt JG, Krieg NR, Sneath PHA, Staley JT, William ST. (1994) Bergey's Manual of Determinative Bacteriology, ninth ed., *Williams and Walkins,* Baltimore. 787.

[266] Felsenstein J. Confidence limits on phylogenies: An approach using the bootstrap, *Evolution*. 1985; 39: 783-791.

[267] Oumer OJ, Abate D. Comparative Studies of Pectinase Production by Bacillus subtilis strain Btk 27 in

Submerged and Solidstate fermentations, *Biomed Res Int.* 2018;1514795.

[268] Tamura K, Nei M, Kumar S. (2004) Prospects for inferring very large phylogenies by using the neighbor-joining method, *Proceeding of the National Academy of Sciences* (USA).101:11030-5.

[269] Singh AK, Mukhopadhyay M (2016) Lipase-catalyzed glycerolysis of olive oil in organic solvent medium: Optimization using response surface methodology. *Korean J Chem Eng* 33(4):1247-1254.

[270] Kiran RRS, Konduri R, Rao GH, Madhu GM (2010) Statistical optimization of endo-polygalacturonase production by overproducing mutants of Aspergillus niger in solid-state fermentation. *J Biochem Tech* 2(2):154-157.

[271] Singh AK, Mukhopadhyay M (2014) Optimization of lipase-catalyzed glycerolysis for mono and diglyceride production using response surface methodology. *Arab J Sci Eng* 39:2463-2474.

www.ingramcontent.com/pod-product-compliance
Lightning Source LLC
Chambersburg PA
CBHW052200220526
45471CB00004B/1745